20 世纪 20 年代中期,外滩码头的繁荣景象　　The Bund in the 1920s

A Stroll along the Bund in the 1930s

You started on its southern end, where it met the Quai de France, and with slow gravity, the massive fronts filed past you. The Asiatic Petroleum Building, on the corner of Avenue Edward VII, led the parade. Next was the Shanghai Club, stodgy and imperial, housing the world's longest bar... You had arrived at Shanghai's most important intersection, the great caesura in the front line of the Bund, Nanking Road. You crossed the street, dodging a dozen rickshaws and one or two tram cars. Your eyes rose along the towering structure of the Sassoon House, Shanghai's tallest building... And on its old luxurious grounds, reaching all the way to the bank of Soochow Creek, there was the British Consulate.

——from Earnest O. Hauser's book *Shanghai: City for Sale* published in 1940

20 世纪 30 年代的外滩　　The Bund in the 1930s

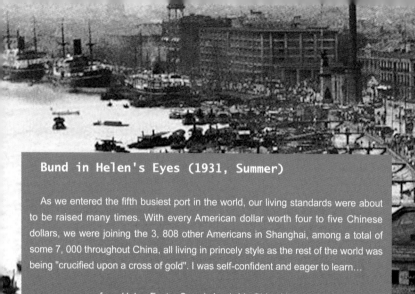

Bund in Helen's Eyes (1931, Summer)

As we entered the fifth busiest port in the world, our living standards were about to be raised many times. With every American dollar worth four to five Chinese dollars, we were joining the 3,808 other Americans in Shanghai, among a total of some 7,000 throughout China, all living in princely style as the rest of the world was being "crucified upon a cross of gold". I was self-confident and eager to learn…

——from Helen Foster Snow's book *My China Years* published in 1984

20 世纪 30 年代的外滩　　The Bund in the 1930s

30 年代漫步外滩

你从黄浦滩的南端洋泾浜走起,就可以看到一所一所的高大房屋。第一家是亚细亚公司,它边上是上海总会,有上海年代最为久远的酒吧。再过去就是日清汽船会社、大英银行、通商银行和轮船招商局……你在电车、汽车和黄包车的中间穿过南京路之后,可以看见那高耸入云的沙逊大厦,门前有华懋饭店的霓虹灯招牌……走到苏州河边时就到了大英领事馆,高踞在一片很大的草地之上。

——摘自美国记者霍赛的《出卖上海滩》(1940)

20 世纪 20 年代的外滩全景　The Panorama of the Bund, 1920s

海伦眼中的外滩 （1931年夏）

当我们来到这世界第五大繁忙的港口时,生活水平顿时提升了好几倍。在这里每个美元可以兑换4到5块中国银元。我加入了上海3 808名(全中国约7 000名)美侨之中,大家都过着王子般的生活。而此时此刻,世界上其他地方的钱袋子都勒得很紧。我充满自信,热切地想要学习……

——摘自海伦·福斯特·斯诺的《我的中国岁月》(1984)

城市行走书系
策划：江岱、姜庆共

上海外滩建筑地图
文字：乔争月
摄影：张雪飞
绘图：孙晓悦

责任编辑：常科实
助理编辑：孙彬
书籍设计：孙晓悦

鸣谢：

郑时龄、吴建中、朱纪华、熊月之、邢建榕、章明、唐玉恩、伍江、常青、钱宗灏、王军、王林、华霞虹、张鹏、林沄、娄承浩、俞挺、邹勋、德尔·罗斯福、米歇尔·嘉娜特、戴仕炳、何成钢、马克·博特伦、陈贤、吕恒中、邓靖、陶震

上海市档案馆
上海市图书馆
上海市政府新闻办公室
上海市旅游局
上海海关
久事集团
上海市徐家汇藏书楼
俄罗斯驻沪总领事馆
新民晚报社
上海日报社
上海市城市建设档案馆
上海市城市规划展示馆
黄浦区人民政府档案馆
上海外滩源投资发展有限公司
上海章明建筑设计事务所

同济大学出版社
TONGJI UNIVERSITY PRESS

CityWalk Series
Curator: Jiang Dai, Jiang Qinggong

Shanghai Bund Architecture
Text: Michelle Qiao
Photograph: Zhang Xuefei
Illustration: Sun Xiaoyue

Editor: Betty Chang
Assistant Editor: Sun Bin
Book Designer: Sun Xiaoyue

Acknowledgements:
Zheng Shiling, Wu Jianzhong, Zhu Jihua,
Xiong Yuezhi, Xin Jianrong, Zhang Ming,
Tang Yu'en, Wu Jiang, Chang Qing, Qian Zonghao,
Wang Jun, Wang Lin, Hua Xiahong, Zhang Peng,
Lin Yun, Lou Chenghao, Yu Ting, Zou Xun,
Tweed Roosevelt, Michelle Garnaut, Dai Shibing,
He Chenggang, Mark Bertram, Chen Xian,
lv Hengzhong, Deng Jing, Tao Zhen

Shanghai Municipal Archives
Shanghai Library
Information Office of Shanghai Municipality
Shanghai Municipal Tourism Administration
Shanghai Customs
Shanghai Jiushi Corporation
The Xujiahui Library
Consulate General of the Russian
　　Federation in Shanghai
Xinmin Evening News
Shanghai Daily
Shanghai Urban Construction Archives
Shanghai Urban Planning Exhibition Center
Huangpu District Archives Bureau
Shanghai Waitanyuan Urban
　　Development Co. Ltd.
Shanghai Zhangming Architectural Design Firm

上海外滩建筑地图
Shanghai Bund Architecture

乔争月 张雪飞 著
Michelle Qiao Zhang Xuefei

同济大学出版社
TONGJI UNIVERSITY PRESS

目录

序言 ·················· 18

外滩拼图 ·················· 25

上海外滩建筑地图 ·················· 30

  外滩信号塔：第一个音符 ·················· 32

  外滩1号：百万美金天际线的起点 ·················· 36

  外滩2号：英伦风情的大班俱乐部 ·················· 40

  外滩3号：公和洋行闪亮登场 ·················· 46

  外滩5号：开启黄金时代 ·················· 52

  外滩6号：通商银行的哥特式摇篮 ·················· 56

  外滩7号：巴洛克穹顶下的电报大楼 ·················· 60

  外滩9号：红与灰 ·················· 64

  外滩12号：外滩的橄榄叶 ·················· 68

  外滩13号：外滩钟楼之巅 ·················· 74

  外滩14号：最年轻的老大楼 ·················· 80

  外滩15号：先锋的古典之作 ·················· 84

  外滩16号：外滩的"希腊神庙" ·················· 88

  外滩17号：大力神托举的报业大楼 ·················· 94

  外滩18号：菲利普的方案 ·················· 100

  外滩19号：外滩的"王宫" ·················· 104

  外滩20号：摩登的"老贵族" ·················· 108

  外滩23号：外滩的斗拱 ·················· 114

  外滩24号：邬达克的外滩工作室 ·················· 118

  外滩26号：保险大楼与外滩风景 ·················· 122

  外滩27号："洋行之王"的上海总部 ·················· 128

  外滩28号：平凡的惊艳 ·················· 132

  外滩29号：追逐"华尔街"的法国银行 ·················· 136

  外滩33号1号楼：英领馆的广玉兰 ·················· 140

  外滩33号2号楼：滨水小楼的迷离身世 ·················· 144

  外白渡桥：这是一座长青桥 ·················· 148

  上海大厦：看见整个外滩的露台 ·················· 152

  浦江饭店：木香萦绕的老饭店 ·················· 156

  俄罗斯联邦驻上海总领事馆：
  反复开关的领事馆 ·················· 162

  黄浦公园：美丽外滩的过去和未来 ·················· 166

拼图外滩 ·················· 174

推荐阅读 ·················· 177

图片来源 ·················· 179

Contents

Preface ··· 18

The Bund Puzzle ································ 25

Shanghai Bund Architecture ············· 30

   The Gutzlaff Signal Tower
   The First Note ······································· 32

   No. 1 - Start of the Billion-dollar Skyline ········ 36

   No. 2 - From Taipan's Club to Waldorf ··········· 40

   No. 3 - Debut of Palmer & Turner ················· 46

   No. 5 - 1920s Grandeur Starts at No. 5 ··········· 52

   No. 6 - Cradle of Chinese Modern Bank ········· 56

   No. 7 - Telegraph Building with Baroque Domes ··· 60

   No. 9 - Red or Gray ································· 64

   No. 12 - An Olive Leaf for Shanghai ·············· 68

   No. 13 - Bell Tower atop the Bund ················ 74

   No. 14 - The Youngest Old Building on the Bund ··· 80

   No. 15 - The Avant-garde, Classical Building ········ 84

   No. 16 - Greek Temple on the Bund ·············· 88

   No. 17 - Press Tower Supported by Atlantes ········ 94

   No. 18 - UNESCO Awarded Building ············ 100

   No. 19 - The "Palace" on the Bund ·············· 104

   No. 20 - A Hotel of Memories ···················· 108

   No. 23 - "Dougong" on the Bund ················ 114

   No. 24 - Hudec's Atelier on the Bund ············ 118

   No. 26 - Monument to Early Preservationist ······ 122

   No. 27 - Headquarter for the "King of Hongs" ··· 128

   No. 28 - Symbols and Refined Details ············ 132

   No. 29 - The Only French Elegance on the Bund ··· 136

   No. 33-1 - Balfour's Vision ························ 140

   No. 33-2
   - The Oldest Surviving Waterfront Building ······· 144

   The Waibaidu Bridge
   - Shanghai's "Grandma Bridge" ··················· 148

   Broadway Mansions Hotel Shanghai
   - The Screen of the Bund ························· 152

   Pujiang Hotel
   - A Hotel Scented with Oriental Woods ··········· 156

   Consulate General of the Russian Federation in Shanghai

   - Openings and Closings ·························· 162

   The Huangpu Park - All for a Better Bund ········ 166

Making a New Bund Puzzle ············· 174

Recommended Readings ················· 177

Image Source ································· 179

序言

随着城市的进步,公众越来越关注历史建筑,文化遗产日期间公众踊跃参观历史建筑的热情令人十分感动。当同济大学出版社"城市行走书系"的又一本新书《上海外滩建筑地图》的书稿放在我的书桌上时,我有良多的感慨:首先是社会各界都在普遍关心建筑,其次是建筑学术界也为普及建筑知识在努力。"城市行走书系"中的《上海里弄文化地图:石库门》《上海邬达克建筑地图》《上海教堂建筑地图》等都深受读者欢迎。上海的外滩建筑更是世界所关注的中心之一,因此这本《上海外滩建筑地图》一定会得到社会公众的喜爱。

外滩是上海的,关于外滩的建筑已经有许多中外文献出版。英文文献以彼得·希巴德的《上海外滩:中国面向西方》(2007),以及丹尼斯·乔治·格罗的《老上海外滩:19世纪的罕见照片》(2012)等最为著称。中文则有常青主编的《都市遗产的保护与再生——聚焦外滩》(2009)、钱宗灏等著的《百年回望——上海外滩建筑与景观的历史变迁》(2005)、薛理勇的《上海外滩》(2012),以及一些其他图片集。这本由乔争月女士等人撰写的《上海外滩建筑地图》则用图文并茂、中英文合璧的方式将外滩建筑呈现给读者。

《上海外滩建筑地图》是一本媒体人的外滩历史建筑寻访和探究的记录,作者有优秀的英文水平,在文献研究方面显然驾轻就熟,作者同时又和建筑界的学者有密切的联系和合作,曾经为《上海日报》和《新民晚报》撰写了许多关于上海历史建筑的报道。她的一些文章收录在《上海西区地标建筑步行导览》中。她也是《上海邬达克建筑地图》的英文作者,并和华霞虹博士一起将意大利学者卢卡·彭切里尼博士和匈牙利学者尤利娅·切侬迪博士的《邬达克》一书译成中文出版。乔争月女士为撰写这本《上海外滩建筑地图》,在徐家汇藏书楼查阅了大量历史文献,并从旧上海的英文报纸《远东评论》《字林西报》《密勒氏评论》《大陆报》中获得了关于外滩历史建筑的许多宝贵资料。同时也做了大量的采访,亲身考察了每一幢建筑,探究其人文内涵。书中附有大量的历史照片和今天的建筑美图,使《上海外滩建筑地图》具有宝贵的建筑历史价值和文化艺术价值。

外滩是东方大都市老上海的标志，也被誉为老上海的心脏和灵魂，还是中国的历史文化街区。外滩之所以闻名世界，一方面是因为这条大道的西侧汇集了上海最具历史文化价值的建筑；另一方面也是因为外滩属于世界最优秀的滨水空间之一。经历了170多年的演变，外滩已经成为上海的标志，代表了上海作为万国建筑博览会的形象。外滩，通常是指从延安东路向北到外白渡桥的一条滨江大道，一般也把沿外滩的建筑包括在外滩的范围内。自延安东路往南的滨江大道称为法租界外滩，但是通常公认的外滩其实并不包括法租界外滩。外滩历史文化风貌区则包括了苏州河北岸的历史建筑在内，因此，本书的内容除了外滩1号以北的23幢近代建筑外，还包括了百老汇大厦、理查饭店和俄罗斯领事馆等。

今天外滩的建筑大多形成于20世纪20年代中晚期及20世纪30年代早期，最早的可以追溯到19世纪60年代，代表了当时世界建筑设计和施工技术的一流水准，属于宝贵的世界文化遗产。外滩的建筑多数都经历过多次重建，可以说一直处于改建和重建的过程中，最终形成今天的外滩城市空间和天际线。外滩的建筑当属近代中国建筑的明珠，日本的《新建筑》杂志在1999年最后一期出版的专辑列举了全世界20世纪的优秀建筑，关于中国建筑只列了两座，就是外滩的原汇丰银行大楼和海关大楼，这也说明外滩在世界建筑史上的重要地位。

由于全书篇幅的限制，作者从多方收集的文献资料只用上了极少一部分，不免感到有一丝遗憾。尽管如此，这本书对认识上海近代建筑，对研究上海近代建筑史的意义是显而易见的。

2015年7月8日

Preface

There has been a growing interest from the general public for Shanghai's historical buildings as the city develops, and I have been deeply moved by the passion to visit the historical buildings during the Cultural Heritage Days. So when *Shanghai Bund Architecture*, a new book of the City Walk Series by Tongji University Publishing House was put on my desk, my immediate thoughts were firstly, the whole society had really shown greater interest and concern for architecture, and secondly, the architecture academia was also endeavoring to spread architectural knowledge to the general public. Previous books of the series, such as *Shanghai Shikumen*, *Shanghai Hudec Architecture* and *Shanghai Church* have all won great popularity among readers. So this new book, with the focus on the buildings of the worldly renowned Shanghai Bund, will certainly be a new favorite of readers.

The Bund makes Shanghai unique. There have been many Chinese and English publications about the Bund architecture, among which the most popular English publications are Peter Hibbard's The Bund Shanghai: *China Faces West* (2007) and Dennis George Grow's *Old Shanghai's Bund: Rare Images from the 19th Century* (2012), and the most popular Chinese publications are Chang Qing's *Urban Heritage Conservation and Regeneration Collection of the Master's Thesis - Focus on the Bund* (2009), Qian Zonghao's *Glancing through a Century - the historical changes of Shanghai Bund Architecture and Sceneries* (2005) and Xue Liyong's *Shanghai Bund* (2012), as well as some other photo collections. This new *Shanghai Bund Architecture* by Ms. Michelle Qiao and her friend will present the Bund architecture with the authors' unique descriptions and vivid photos, in both Chinese and English.

Shanghai Bund Architecture contains the actual records of Ms. Qiao's research and exploration along the Bund from her journalistic point of view. The author's excellent English allows her to research the precious historical English archives. She also has close relationships and collaborations with architectural scholars. She is a columnist on the city's historical buildings for Shanghai Daily and Xinmin Evening news. Some of her works have been collected into the book *A Walker's Guide to Old Western Landmarks in Shanghai* (2008). She is also author for the English version of the book *Shanghai Hudec Architecture* and translated the book *Laszlo Hudec* by Italian scholar Luca Poncellini and Hungarian scholar Júlia Csejdy in cooperation with Dr. Hua Xiahong.

To attain precious first-hand material about the Bund historical buildings, Ms. Qiao has done arduous research in Xujiahui Library where abundant historical archives from Shanghai's old English newspapers are kept, such as *Far Eastern Review*, *North China Daily News*, *Milliard's Review* and *The China Press* etc. For every building, she has also done

many interviews and field studies so as to get a comprehensive understanding. And many beautiful historical and present photos selected from this research process are another worthy effort to display the architecture's historical, cultural, and artistic values.

More than the symbol of old Shanghai, the Bund has been acclaimed as the heart and soul of old Shanghai, the oriental metropolis. It is now a national historical cultural street. The international popularity of the Bund mainly comes from two aspects. One is that the western side of this promenade is the congregation of historical buildings with the most historical and cultural values in Shanghai. The other is that the Bund area is now one of the best waterfront spaces in the world. After more than 170 years' evolution and development, the Bund has become the symbol of Shanghai and a typical representative of the city's image as an exhibition of world architecture.

Usually the Bund indicated the promenade starting from Yan'an Road E., stretching northward until the Waibaidu Bridge. And in general it also includes the architecture along the promenade. The southern part of it, which spreads from Yan'an Road E. southward in the former French concession, could be called the French Bund, but the French part has not been publically recognized as part of the Bund yet. The official Bund historical area also includes the historical buildings on the north bank of Suzhou Creek. Therefore, this book introduces not only the 23 waterfront buildings starting northward from No. 1, but also buildings on the north bank, such as the Broadway Mansions, Astor House and the Russian Consulate, etc.

Most of the Bund architecture that we see today came into shape from the middle or late 1920s to the early 1930s, and the earliest could be traced back to the 1860s. They all have demonstrated the most popular western architectural design and first-class construction technology of their times. Most Bund buildings have been kept renovation and reconstruction over the years, which eventually shaped the urban space and skyline of the Bund today. Thus undoubtedly the Bund architecture is a glistening pearl of Chinese modern architecture. Japanese magazine Shinkenchiku Magazine listed the most excellent 20th century architecture in the world in its last special edition of 1999. Only two buildings from China, the former HSBC building and the Custom House, were listed. And both are from the Bund. This fully shows the important status of the Shanghai Bund architecture in the world's architectural history.

Because of the book's length limit, only a small selected portion of the author's archival collection have been used, which is a little bit pity. However, this book is obviously meaningful to people to know about Shanghai's modern architecture and to do study about Shanghai's modern architectural history.

Zheng Shiling

July 8, 2015

外滩拼图

今日的外滩摩登、开阔、迷人,但许多人不知道外滩的发展其实是个意外,把外滩划为租界居然也是清政府上海道台的"创意"。

1843年上海开埠后的首任道台宫慕久,是位深受儒家思想熏陶的传统男人,一心想着把华洋隔离开来。他从农耕文明的角度考虑,位于上海县城外的黄浦滩土壤贫瘠、蚊虫甚多,耕种居住两不宜,把"洋鬼子"集中到此地管理再好不过。

巧的是首任英国驻沪总领事巴富尔也相中了外滩。他从海洋文明的角度审视这片泥滩,是一块地理位置绝佳的经商宝地,只要能够赚钱,土壤和蚊子都不是问题。道台和领事从不同角度思考,却将眼光同时投向了外滩,一拍即合。就这样,上海的第一个"自贸区"诞生了。

开埠后短短数年,这片泥滩就奇迹般地冒出了一座座洋行,逐渐升起一座迷你的欧洲城市。外滩地区现存的建筑风貌被称为"外滩三期",主要成形于20世纪20年代的"黄金年代"。

仅有一公里多长的外滩黄金段既是重要码头,又有绝佳江景,在这里拥有一座建筑来代表自己,是彼时活跃沪中的各种势力梦寐以求的。除中国业主外,外滩大楼的业主来自当时在华势力最强的几个国家,美、英、法、日、俄和德(德国总会,后拆除),一个都没有少。其所代表的行业也是旧上海滩利润最高的,包括早期靠鸦片起家的老牌贸易洋行、贸易支柱产业航运业、后来增长强劲的金融业和租界人口剧增带来的暴利的房地产业。

这片中国最大的近代建筑群涵盖了19世纪殖民地风格的外廊建筑、唱主角的新古典主义大厦、增色不少的哥特复兴小楼和与美国同步流行的装饰艺术派大楼,几乎是一部生动的缩略版近现代西方建筑史。而在这铅灰色西洋建筑群中,也不乏中国元素灵动的影子,如16号柱头上的美丽回纹,以及23号中国银行的云纹与石质斗拱。

外滩的传奇多,但最打动我的是两位美国女记者的外滩故事。

一位是后来成为斯诺夫人的海伦。她1931年到沪当天就结识了《西行漫记》的作者——美国记者斯诺,第二年嫁给了他。后来,这位美领馆女秘书华丽转身为著名新闻人。海伦下榻在外滩礼查饭店,和斯诺在南京路巧克力店一见钟情,应允求婚也在外滩。

另一位更传奇的是《宋氏三姐妹》作者项美丽。1935年，这个爱冒险的姑娘失恋又失业，陪姐姐到上海游玩散心，却被这座城市迷住了，住了整整六年。在上海，她不仅成为犹太富商沙逊的密友，还嫁给出身名门的新月派诗人邵洵美为妾，为日后采访宋家姐妹打下重要基础。项美丽工作、生活和娱乐都在外滩。

两位原本平凡的美国女孩，在"魔都"上海，在气场强大的外滩，她们的人生都发生了如此奇妙的逆转。这激起了身为同行的我探访外滩魔力场的兴趣，决心借用一座座历史建筑完成迷人的外滩拼图。

我探索外滩的角度略有特别，研究资料主要源于近代英文报纸。上海是一座先有英文报纸后有中文报纸的城市，第一份报纸是创刊于1851年的英文《北华捷报》。而几乎每座外滩建筑奠基或落成时，沪上主流英文报纸都会派一位我百年前的同行，去实地采访并撰写图文并茂的深度报道。他们的新闻作品就是非常珍贵的关于外滩建筑和上海城市历史的一手档案资料，为我描绘出一个原汁原味的外滩。

在各路"神仙"的帮助下，我幸运地敲开了几乎所有外滩建筑的大门，得以探访外滩大楼神秘的内部空间，同时邀请到建筑师好友张雪飞从专业角度来捕捉外滩建筑之美。

历史钩沉加上实地采访，来来回回，常常发现一个鲜为人知的故事；登上一个云朗风清红旗飘展的外滩露台；抬头看见一条绝美的纹饰带，都让我激动不已。外滩的内涵更胜其外表，那么精彩，那么丰富。

经过近三年的努力，我终于把外滩地区最知名的43幢历史建筑逐个写了一遍，分别以中英文专栏的形式在《新民晚报》和《上海日报》刊登。

很多中外读者来信希望我带领大家逛逛外滩，这也给予我很大的写作动力。鉴于篇幅有限，本书共选取了30幢外滩近代建筑，以洋泾浜（今延安东路）为界，集中在旧"公共租界"外滩的一段。

今年4月，住建部和国家文物局公布了首批30个中国历史文化街区，外滩成为上海唯一入选的街区。希望这本媒体人的历史建筑探寻笔记，能陪伴大家探索这气场强大的外滩。

乔争月
2015年5月

The Bund Puzzle

The bund today is open, modern, and fascinating, but few of the thousands of visitors everyday are aware that the development of the Bund was quite accidental.

It was Shanghai Taotai (Qing's governor)'s idea to allocate the Bund area as the city's first British settlement. Gong Mujiu, the first Taotai since Shanghai was forced to be an open port in 1843, a traditional Chinese under heavy Confucius influence, decided to seclude the foreign settlers from the locals to avoid troubles with foreign forces. From the traditional Chinese agricultural perspective, which was land-based, infertile soil and heaps of mosquitos made the Bund area hard for planting or dwelling. It was also outside the Shanghai County, a seemingly great place to allocate the "foreign devils."

Meanwhile, George Balfour, the first British consul to Shanghai, and his colleagues also cast their eyes on the Bund area. Judging from the western commercial perspective evolved from the Mediterranean marine civilization, they were amazed at the strategic location and unlimited potentials of this mud land. The westerners came to Shanghai for trade and business, so infertility and mosquito were not of any problem.

The 1843 map of the Bund was crisscrossed with farm fields, graveyards, and even a rooster fighting yard. Only six years later dozens of "hongs," the earliest foreign trade companies mushroomed and a mini European city took the shape. The city's first "free-trade pilot zone" came into being.

As Chinese art master Feng Zikai (1898–1975) said, architecture was powerful propaganda owing to its enormous size and closeness to daily life. The Bund became a huge gallery for showcasing the powers.

Almost every foreign power active in old Shanghai had buildings on the Bund, including Britain, France, America, Japan, Russia, and Germany. The owners of the Bund buildings were representatives of the most important, profitable businesses, ranging from trade, shipping, banking to real estate.

Today the Bund has preserved not only China's largest modern western architectural congregation, but also the largest in the Far East Asia as well, exhibiting a rainbow of architectural styles including 19th century colonial veranda, neoclassical, Gothic-revival and Art Deco styles, which is almost like a living, condensed exhibition of modern western architectural history.

And some Chinese elements twinkle among the line of grey western buildings, such as cloud patterns and stone Tougongs (traditional Chinese brackets) all over the Bank of China at No. 23.

The bund has always been legendary, but as to what inspired me most were the Bund

adventures of two American women journalists.

One was Helen Foster Snow, who spent her first Shanghai night at the Astor House Hotel on the Bund in 1931. She got to know *Red Stars over China* author Edgar Snow very soon, became Mrs. Snow the next year and thus started a career as a renowned journalist herself. They met in a chocolate shop on Nanking Road and Edgar proposed on the Bund.

The other was Emily Hahn, who authored the best-selling "Soong Sisters." This adventurous American girl's scheduled Shanghai holiday in the 1930s prolonged from weeks to six years. She married a Chinese poet from a noble family as his concubine, became a close friend of Jewish tycoon Victor Sassoon and wrote fantastic stories about China for the *New Yorker*. She worked, lived and enjoyed her life all around the Bund.

They aroused my curiosity, how two ordinary American girls, one a secretary and the other jobless, changed their lives dramatically on the powerful Bund in this magical city of Shanghai. What were the stories hidden behind each of the buildings, other than those brief introductions on a guidebook? So I decided to write about the beautiful bund buildings as playing an amazing puzzle.

The very special source of my Bund research was the old English newspapers. For almost every building on the Bund when it was laid the foundation stone or just completed nearly a century ago, a journalist was assigned to interview the building and write an in-depth feature story. Their writings became precious archives today and painted a rather authentic picture of the Bund for me.

Earlier this year, the Bund was included on a new list of the country's historical streets. It is Shanghai's only street on the list of 30 streets announced by the State Bureau of Cultural Relics, which is another new evidence of lasting passion and love for the Bund.

During the last three years, I visited the Bund almost every week. With the help of various organizations and friends, I was fortunate to have an in-depth exploration of almost all buildings on the Bund. My friend Zhang Xuefei has helped to capture the architectural beauty of the Bund from the eyes of a professional architect.

Appealing to both readers from home and abroad, I ran the Bund column simultaneously in the city's two leading English and Chinese newspapers. Interestingly among the feedbacks, many have asked if I can arrange a walk tour along the Bund. So I've picked 30 Bund buildings in the former International Settlement for the book. With this "Shanghai Bund Architecture", a journalist's exploration records of the Bund, I sincerely wish it can accompany you for me, to explore the magical Bund, from No. 1 to No. 33.

Michelle Qiao
May 2015

外滩信号塔	外滩 23 号	The Gutzlaff Signal Tower	No. 23
中山东二路 1 号甲	中山东一路 23 号	1A Zhongshan Rd. (E-2)	23 Zhongshan Rd. (E-1)
外滩 1 号	外滩 24 号	No. 1	No. 24
中山东一路 1 号	中山东一路 24 号	1 Zhongshan Rd. (E-1)	24 Zhongshan Rd. (E-1)
外滩 2 号	外滩 26 号	No. 2	No. 26
中山东一路 2 号	中山东一路 26 号	2 Zhongshan Rd. (E-1)	26 Zhongshan Rd. (E-1)
外滩 3 号	外滩 27 号	No. 3	No. 27
中山东一路 4 号	中山东一路 27 号	4 Zhongshan Rd. (E-1)	27 Zhongshan Rd. (E-1)
外滩 5 号	外滩 28 号	No. 5	No. 28
中山东一路 5 号	中山东一路 28 号	5 Zhongshan Rd. (E-1)	28 Zhongshan Rd. (E-1)
外滩 6 号	外滩 29 号	No. 6	No. 29
中山东一路 6 号	中山东一路 29 号	6 Zhongshan Rd. (E-1)	29 Zhongshan Rd. (E-1)
外滩 7 号	外滩 33 号 1 号楼	No. 7	No. 33-1
中山东一路 7 号	中山东一路 33 号	7 Zhongshan Rd. (E-1)	33 Zhongshan Rd. (E-1)
外滩 9 号	外滩 33 号 2 号楼	No. 9	No. 33-2
中山东一路 9 号	中山东一路 33 号	9 Zhongshan Rd. (E-1)	33 Zhongshan Rd. (E-1)
外滩 12 号	外白渡桥	No. 12	The Waibaidu Bridge
中山东一路 12 号	横跨苏州河下游河口	12 Zhongshan Rd. (E-1)	Above the confluence of
	东临黄浦江		Suzhou Creek and
外滩 13 号		No. 13	Huangpu River
中山东一路 13 号	上海大厦	13 Zhongshan Rd. (E-1)	
	北苏州路 20 号		
外滩 14 号		No. 14	Broadway Mansions Hotel
中山东一路 14 号	浦江饭店	14 Zhongshan Rd. (E-1)	Shanghai
	黄浦路 15 号		20 Beisuzhou Rd.
外滩 15 号		No. 15	
中山东一路 15 号	俄罗斯联邦驻上海总领事馆	15 Zhongshan Rd. (E-1)	Pujiang Hotel
	黄浦路 20 号		15 Huangpu Rd.
外滩 16 号		No. 16	
中山东一路 16 号	黄浦公园	16 Zhongshan Rd. (E-1)	Consulate General of the
	中山东一路 500 号		Russian Federation in Shanghai
外滩 17 号		No. 17	20 Huangpu Rd.
中山东一路 17 号		17 Zhongshan Rd. (E-1)	
外滩 18 号		No. 18	The Huangpu Park
中山东一路 18 号		18 Zhongshan Rd. (E-1)	500 Zhongshan Rd. (E-1)
外滩 19 号		No. 19	
中山东一路 19 号		19 Zhongshan Rd. (E-1)	
外滩 20 号	提示：不同颜色对应区域详	No. 20	Tips: please refer to the map on the
中山东一路 20 号	见右侧建筑地图	20 Zhongshan Rd (E-1)	right side for the colored patches.

上海市中心地图

北苏州路	Beisuzhou Rd.
苏州河	Suzhou Creek
大名路	Daming Rd.
南苏州路	Nansuzhou Rd.
外白渡桥	The Waibaidu Bridge
圆明园路	Yuanmingyuan Rd.
黄浦公园	The Huangpu Park
北京东路	Beijing Rd. (E)
滇池路	Dianchi Rd.
天津路	Tianjin Rd.
南京东路	Nanjing Rd. (E)
黄浦江	Huangpu River
河南中路	Middle Henan Rd.
九江路	Jiujiang Rd.
南京东路（地铁站）	East Nanjing Rd.
汉口路	Hankou Rd.
中山东一路	Zhongshan Rd. (E-1)
福建中路	Middle Fujian Rd.
福州路	Fuzhou Rd.
广东路	Guangdong Rd.
延安东路	Yan'an Rd. (E)
中山东二路	Zhongshan Rd. (E-2)
人民路	Renmin Rd.

本图为位置示意，与实际比例不符
Illustration is not proportional to the actual scale

轨道交通及车站
Metro Stations

外滩信号塔
The Gutzlaff Signal Tower

第一个音符
The First Note

外滩从哪里开始？也许，白色的信号塔就是答案。

信号塔初建于1884年，兼有报时和气象预报两个功能，主要负责"广播"由徐家汇天文台传来的信息。在没有无线广播的年代，信号塔模仿伦敦格林尼治天文台的做法，用落球的方式来报时。

1924年出版的由陈伯熙所著的《上海轶事大观》中提到，信号塔的黑色大铁球于每天11点45分升到半旗杆的位置，5分钟后升到顶端，11点55分第一次下坠，然后再升至顶端，最后于正午准点再次下坠。黄浦江上的船员们便会据此校正航海钟，以确保航行安全。

这个方法是格林尼治天文台于1833年首创的，后来各港口城市纷纷效仿。

研究徐家汇天文台的历史学家王钱国忠认为，外滩信号塔的报时方法很科学。因为塔身很高，在上海市区的许多地方都可以看到它，能够直观地向市民预告雷电台风。虽然这个方法很原始，但是人们可以马上看到信号。在当年，信号塔不仅让上海市民的生活更方便，对于黄浦江上外国军舰的出航也有很大帮助，所以曾得到法国海军司令的褒奖。

信号塔预告气象的方法也十分科学。定时悬挂彩色信号旗以进行气象预告，这些形状和颜色各异的信号旗分别表示大风、台风、雾和风向等不同的气象信息。1908年版的《上海指南》中刊登了详细的信号旗信息图。

1906年的夏天，信号塔的木桅杆被台风折断，子午球坠落。这次事件促使法租界公董局于1907年在原址重建了一座坚固的钢筋水泥信号塔。塔身高近50米，1927年扩建了裙房，这就是今日外滩信号台的全貌。

1993年外滩道路拓宽时，信号台整体向东南移位20余米，但历史旧貌得以完好保留。如今其一层改造为展示厅，二层是为城市规划展示馆服务的会议室。一座陡峭的螺旋形铁梯通往曾高悬铁球和彩旗的信号塔顶端。

参观指南

一层计划作为展览空间对公众开放，但尚未宣布开放时间。可以关注这座建筑在外滩空间的重要位置。

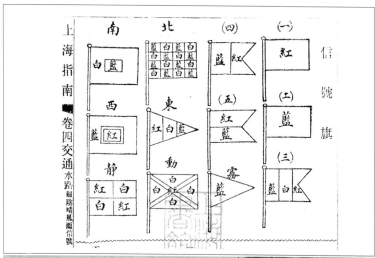

上图：信号旗信息图（《上海指南》，1908）下图：20世纪10年代外滩信号塔旧貌
Above: Meteorllogical flags on *Shanghai Guide*, 1908　Below: A historical photo of the Gutzlaff Signal Tower

"The Gutzlaff Signal Tower is like the first note of the Bund symphony," says Tongji University Vice President Wu Jiang.

The lovely white tower, which has recently been restored, was originally erected in 1884 in the former French Bund to transmit weather forecast from the Xujiahui Observatory to the public.

The latter, founded in 1872 by the French, was one of the most famous observatories in the Far East, and is still in use today.

Perched at the crossroads of the Bund and Yang Ching Bang (today's Yan'an Road E.), the white tower was also used to tell the time. Initially it reported time only once daily at noon, but in 1909 it began using flash signals to indicate it was 9pm. It began using a new technique to report the time after 1914.

According to Chen Boxi's 1924 book *A Comprehensive View of Shanghai Anec-dotes*, the big iron ball would be lifted to half the height of the pole at 11:45am, to the peak at 11:50am, and fall for the first time at 11:55am. It was then lifted to the peak again and fell at exactly noon. In this way, sailors would prepare and adjust their navigation clocks to the correct time.

This advanced method was first used by London's Greenwich Observatory in 1833 and was copied by Japan's Yokohama, Kobe, and other port cities.

The tower had also used meteorological flags to report the wind speed and direction. The 1908 version of *Shanghai Guide* even published a complete set of meteorological flags in all shapes and colors and their implications.

Historian Wangqian Guozhong admires the tower as a scientific, direct way to spread information regarding weather and time, although ancient in today's world. The tower on the Bund was so high that people in most parts of downtown Shanghai could see it easily.

In the summer of 1906, the pole of the old wooden tower broke during a strong typhoon. The Xujiahui Observatory used bricks and reinforced concrete to repair it. But one year later the tower was rebuilt as a 50-meter-high concrete-and-steel structure.

In 1927, a podium was added alongside the tower. Until then the tower was almost identical to its appearance today.

In 1993, the signal tower was moved 20 meters southeast when the Bund was renovated. Today the ground floor is an exhibition room while the second floor serves as a meeting room that is used by the Shanghai City Planning Exhibition Hall.

Professor Wu says the perfect silhouette of the 1.5-kilomter-long Bund is like a beautiful symphony, with the former HSBC Building (No. 12) and the Customs House (No. 13) as the first climax and the Fairmont Peace Hotel (No. 20) and the Bank of China (No. 23) as the second.

"The nice ending is Broadway Mansion fronting the Huangpu River," he adds.

Well, all starts from the first note.

Tips

It has a plan to open the ground floor to the public, which is not announced yet. Please note its important location of the Bund region.

外滩 1 号 No.1

百万美金天际线的起点
Start of the Billion-dollar Skyline

亚细亚大楼的地址十分显赫，外滩更著名的大楼都无法与之相比。它在1916年初建时是当之无愧的外滩第一楼，不仅体量最大，还是"百万美金天际线"的起点。

大楼由英国商人麦边所建，位于外滩和爱多亚路（今延安东路）的转角。大楼落成一年多前，爱多亚路还是淤塞严重的洋泾浜。填河筑路后周边环境大为改善，为大楼吸引了不少租户。

外滩1号是一座"回"字形的巨型写字楼，立面为中规中矩的经典构图，在朝向延安东路和外滩的两个立面，分别设计有两行古希腊爱奥尼立柱，柱头涡卷优美，是点睛之笔。

电灯发明以前，沪上人家多用火油或蜡烛照明。美国记者豪瑟1940年出版的名著《买卖上海滩》中提到，中国传统的油灯换用火油后更明亮。20世纪初，中国每年要消费650万桶火油，主要由英国壳牌旗下的亚细亚和洛克菲勒的美孚公司控制。

"外滩1号就是亚细亚火油公司的办公室。除了石油，该公司还销售润滑油、蜡烛和其他技术类产品。公司有工程部、统计和广告部、船运部和工作车间，甚至有自己的码头。"豪瑟写道。由于租用大楼的亚细亚公司名气大，人们后来习惯性地称这座建筑为"亚细亚大楼"。

洋油公司采取低价倾销的策略，占领中国市场相当成功。到1949年，亚细亚在中国已有7 000多名雇员，公司把标志骄傲地挂到外滩1号的正门上。

亚细亚公司在中国经营活动持续到1950年左右，之后大楼由上海机械设计研究院和上海丝绸公司等单位使用。

如今外滩1号神秘谢客，从半遮半掩的大门望进去是满布尘土的工地，旧貌损坏严重。

所幸从黄浦区档案中的一段文字，我们仍可以想象昔日外滩第一楼内部的模样。

"楼层高敞，分间较大，室内地板，过道宽阔，马赛克地坪。楼内全部用威克公司承造硬木转门。天井四壁和过道墙面，均贴白色瓷砖，走廊窗高2米，晴天楼宇十分明亮。"

参观指南

建筑仍在改造中，不对外开放。请注意建筑所在的重要位置，南侧的延安东路曾经是英法租界的分界线。

1915年，爱多亚路（今延安东路）路口街景　Avenue Eduard VII (today's Yan'an Road E.) in 1915

No. 1 on the Bund, the Asiatic Petroleum Co Building, has a prominent address that none of the more famous Bund buildings would have.

Back in 1916 when the edifice was just erected, it was called the No. 1 Building on the Bund, not only because of the address, but also due to its stunning scale. Covering more than 10,000 square meters, it was indeed the grandest building on the Bund of its time, a remarkable start of the "billion-dollar skyline."

No. 1 sits in a very eye-catching location, beautifully spreading on the corner of the newly constructed Avenue Eduard VII (today's Yan'an Road E.) and the Bund.

Just more than one year before the building was completed, Yang Ching Bang was still a small river heavily retarded by silt that divided the former French concession and the former British Settlement. The surrounding area had improved dramatically after the river was filled up and turned into a road (Avenue Eduard VII). That attracted numerous tenants for the new office building.

Though initially erected by British merchant George McBain, the building is widely known today for one of its earliest occupants, the Asiatic Petroleum Co, a branch of Shell Petroleum, which sold kerosene and candles that Chinese used for lighting before electric lights were invented.

According to American correspondent Earnest O. Hauser's 1940 book *Shanghai: City for Sale*, the lamps of China burned brighter with kerosene, and early last century China consumed 6.5 million barrels of oil every year. Shell's Asiatic Petroleum Co shared the control of the market with Rockefeller's Standard Oil Chinese division known as "Mei Foo."

"At No. 1 on the Bund were the offices of the Asiatic Petroleum Co, the Oriental sales division of the powerful Anglo-Dutch Shell group. Besides its gasoline department, it had an important aviation section, sections for the sale of lubricants, candles and technical products. It had its own engineering department, its statistical and advertising divisions, its shipping department and its workshop. It owned wharves and had a large 'floating staff.' Its compradore was Mr Dow Ding Yao," Hauser wrote.

Before 1949 the company had more than 7,000 employees all around China and had put its logo on the facade of the No. 1 building.

The Asiatic Petroleum Co continued its operation in China until the 1950s, after which the building became the office for state-owned organizations, including the Shanghai Metallurgical Design & Research Institute and the Shanghai Silk Co. The former residence for Asiatic Petroleum's high-ranked employees, a rainbow of garden villas on Julu Road, has been renovated into a boutique hotel.

20世纪20年代的亚细亚火油公司大楼
The Asiatic Petroleum Co Building, 1920s

Today the former No. 1 building on the Bund is mysteriously empty, one of the few empty buildings on the Bund waterfront. Looking through the half-closed grand gate, an unfinished renovation project is visible.

Fortunately, a paragraph in the Huangpu District Archives describes the interior of No. 1:

"The ceiling was high, the open room was spacious and the corridor was wide. The floor was paved with mosaics. Some corridor walls were decorated with white ceramic tiles. All the gates were equipped with hardwood revolving doors. The windows soared an astonishing two meters high, so the rooms were all very sunny and bright."

Tips

Pay attention to the important location of the building near the border of the former International Settlement and the former French concession; You can also visit the former residence for APC employees, a rainbow of garden villas at 889 Julu Rd, renovated into a boutique hotel, Julu Garden.

外滩 2 号 No.2

英伦风情的大班俱乐部
From Taipan's Club to Waldorf

外滩 2 号没有穹顶也不高大,但风格优雅,低调地栖身于中山东一路的南端。钢筋混凝土结构,外立面由石材铺就,比例匀称。一眼望去是 6 根爱奥尼石柱排成的柱廊,引人注目。

每个外滩建筑都有故事,但论及身份的跌宕变迁,没有谁可与外滩 2 号——原上海总会相比。

上海总会成立于 1861 年,会员全为男性。据 1934 年版《上海导览》介绍,加入总会虽难,但国籍并不局限于英国。

1911 年落成的大楼提供了英国绅士需要享受的一切,包括著名的"L"形长吧、弹子房、吸烟室、江景绝佳的餐厅和阅览室。设施设计很人性化,阅览室大木桌中央竟有类似乐谱架的阅读架,便于绅士们轻松翻阅大部头的著作。

在旧时的报纸上,上海总会是被频繁提及的知名场所,人们称这里是"午餐时间遇见所有英国大班的地方"。当年公共租界的许多要事,都是大班们在总会的房间里边喝威士忌边讨论的。

"虽然上海总会太英式、太保守了，而且要成为会员非常困难，但是到了上海如果不应某位会员邀请，到总会去喝杯午餐前的苦味杜松子酒，这趟上海之行就不算圆满。"英国记者巴伯在《上海的陷落》一书中写道。

美国女记者库恩曾撰文回忆总会著名的"长吧"，"这条超过100英尺长、由抛光的桃心木制成的长吧，据说是世界上最长的吧台。透过酒吧的凸窗可以看到黄浦江忙碌的港口。长吧的好座位都留给扬子江的引航员们，因为是他们驾驶船只巧妙地绕过暗流和浅滩，带领人们从入海口来到上海。"

据说长吧也有"潜规则"。大班们总爱聚集在吧台临窗的一侧，靠近位置最好的引航员们。而初来乍到的新会员只能坐进另一侧的昏暗角落里玩骰子。如果后来新会员的地位有所提升，那么他在长吧的位置也可以相应挪动。

太平洋战争爆发后总会命运多舛，1941年被侵华日军占领，1949年后由中国百货公司华东采购站等单位使用。1956年海员俱乐部在这里开张，成为城中少数几个可以吃到罗宋汤的地方，后又化身著名的东风饭店。1988年洋快餐进军上海，拆去长吧，把昔日深色优雅的大班俱乐部改成色调轻浅的餐饮环境，开了上海第一家肯德基。

2011年外滩2号在百岁生日之际，又历经一次整修，复原了长吧，变身奢华的华尔道夫酒店。

这次谨慎的"手术"后，大楼许多地方恢复了历史旧貌，就像1912年英文远东时报记者描述的那样："室内设计美轮美奂。宽阔的花岗石楼梯通向华丽的大厅；高达17英尺9英寸（约5.4米）的爱奥尼柱从精美的黑白大理石地坪上拔地而起；瑰丽的石柱支持着楣构和拱券；回廊在柱间向厅内作圆弧形出挑；大厅的顶部是拱形的玻璃天棚。"

长吧也根据历史照片原样修复。旧照中乳白色的小桌和靠背椅换成了巧克力色的咖啡桌和真皮沙发，与深棕色吧台倒也搭配，长吧的风貌不减当年。如今外滩2号矗立在黄浦滩已整整104年，低调而奢华，浸满了历史与回忆。

参观指南

建议到按照历史旧貌复制的长吧坐坐，喝一杯，在大班时代的怀旧气息里沉浸一会。酒店的电梯和壁炉都是原物。

上海总会门前街景
Street view fronting the Shanghai Club

Every building on the Bund has a story to tell, but none of them has had such an array of important and interesting roles as the Shanghai Club.

Without domes or peaked roofs, the elegant edifice modestly perches on the south end of the riverfront avenue. The facade is carried out in ornamental stone, relieved by six granite columns and plinth. The walls are of steel-reinforced concrete. Men in black now guard the gate of the structure with a giant glass rain roof.

The club had provided almost everything that a British gentleman needed to enjoy, including the L-shaped, 100-foot-long bar, longest in the East at one time.

There was also a billiards room with raftered ceiling and leaded windows, a smoking room, a dining salon overlooking the river, a cards room and reading rooms.

Archive photos show spacious wooden tables in the library, which was even equipped with reading shelves in the center that resembled music stands to make it easier to read big-size books. On the second and third floors were 40 bedrooms, each with a full bathroom.

In memoirs about Shanghai in the 1920s, the Shanghai Club frequently appears and is often vividly described. It was known as "a place where you would probably meet all the taipans of the British community at lunch time."

"However stuffy and 'British' the club, however difficult it was to become a member, no visit to Shanghai was complete without an invitation by a member for a pre-lunch pink gin," wrote Noel Barber in the book *The Fall of Shanghai*.

Many things, including the founding of the Hong Kong and Shanghai Banking Corporation, are said to have been discussed over a glass of whisky at the Shanghai Club.

Irene Corbally Kuhn, a foreign correspondent in Shanghai in the 1920s, recounted stories of the renowned long bar in "Shanghai: The Way It Was" in 1986.

"More than 100 feet of dark, polished mahogany, it was said to be the longest bar in the world. A wide bay window in the barroom overlooked the frenzied harbor traffic. Tables there commonly were reserved for that colorful breed, the Yangtze River pilots, the men who negotiated the tricky passage through shoals and sand bars from the estuary to

Shanghai and beyond."

In an almost rigid, mysterious protocol, the taipans congregated by custom at the window end of the bar near the pilots while new comers played liar dice in the dark recesses at the far end. After a period of time a member would move up the bar as his importance increased.

As Shanghai's golden 1920s and early 1930s slipped away, the fate of Shanghai Club began to change. It was occupied by the Japanese army in 1941, reopened in 1945, closed again in the summer of 1949 and opened by the Shanghai Department Store Co.

In 1956 the former Shanghai Club was opened again as a club for international sailors, one of the few places that offered Western food and borscht. Only 15 years later, the building was converted to the famous Dong Feng Restaurant, which was remembered fondly by many locals since dining out during that period was rare.

In 1988 KFC demolished the long bar and renovated it into Shanghai's first KFC restaurant. The dark-toned long bar room became the light-colored KFC restaurant. The former Shanghai Club famous for its steak-and-kidney pies as luncheon began serving fried chicken.

Fortunately when the building celebrated its 100th anniversary in 2011, it was reopened as the luxurious Waldorf Astoria Hotel.

After careful "surgery" and "make-up," the building was revived and looked almost as it was described in the English-language newspaper *Far Eastern Review* in 1912.

"The interior of the Club is sumptuous.

昔日长吧 The long bar

A broad flight of granite steps leads to a magnificent hall 90 feet in length, 39 in breadth and 41 feet 6 inches in height. Rising from a handsomely designed floor of black and white marble are magnificent Ionic columns 17 feet 9 inches high supporting entablatures and arches and surmounted by a heavy dentiled cornice and a barreled ceiling of glass."

Though lacking the original blueprints, architects reproduced the Long Bar based on archival photos. The reproduction looks authentic, though a row of original white tables and white chairs have been replaced with brown coffee tables and leather couches. Even the original black ceiling fans were reproduced. And finally the former Taipan's Club fills a proper role.

Tips

Having a drink at the long bar is a nice way to soak up some memories; the decor is nostalgic. Note the original elevator and fireplace.

外滩 3 号 No.3
公和洋行闪亮登场
Debut of Palmer & Turner

　　1912年，一位年轻建筑师受公司委托，千里迢迢从香港赴上海开拓市场。实力不俗的他运气很好，首个项目就一炮打响。公司由此打开局面，后来成为上海滩最叱咤风云的建筑事务所。外滩现存的23幢临江历史建筑中有9幢是他们的作品。

　　这位建筑师名叫乔治·威尔逊，当年只有32岁，替远东地区大名鼎鼎的公和洋行工作。而这个幸运项目就是外滩4号——联保大楼，2001年开始改造后被叫作"外滩3号"。大楼1916年落成时被英文《远东时报》称为最值得关注的建筑设计进步之一。这幢体量巨大的六层建筑使用了当时最先进的钢框骨架，风格为自由复兴式，垂直的线条勾勒出骨架结构。

　　"建筑的线条十分美丽，结构也优雅到了极致。对于公和洋行的建筑师来说，实在是一座值得纪念的丰碑。"报道如此称赞。

　　"外滩3号"是外滩近年第一座全

面商业改造的建筑,但由于对历史旧貌改动大而引起争议,2004年重新开张后成为时尚地标,设有数家顶级餐厅、奢侈品专卖店和艺术画廊,受到高端消费者和商家追捧。但历史保护界认为该项目仿佛打开了潘多拉的盒子,把横流的物欲带回外滩。

回到一个世纪前,大楼的一到五层曾出租作为办公室使用,顶层是配有屋顶花园的高档公寓。这个高度既能享受江风,又隔绝了外滩路面的尘嚣,在申城湿热的夏天十分舒适。建筑师最会挑地方,1918年曾为燕京大学(今北京大学)做规划的美国建筑师茂飞就把公寓和事务所都设在顶楼,认为这是城中最棒的办公楼。

英商有利银行曾长期租用底层营业,所以大楼又名"有利大楼"。1949年后由上海市民用建筑设计院使用。

据1916年报道,底楼连续券拱的柱础是花岗石的,支承着假石贴面的砖墙。主入口的两侧竖立巨大的花岗石柱,上面是有雕饰的山形墙,将来客引向壮观的门厅。这幢大楼今天也耐看,真是公和洋行上海滩的开山之作。

公和洋行1868年创办于香港,当时中文名叫"巴马丹拿"。联保大楼建成后,公司在楼内设办事处,并根据上海文化取了"公和洋行"这个好名字。

年轻有为的威尔逊几年后当上合伙人,带领公和洋行在"远东巴黎"大展宏图,陆续完成外滩天际线上最壮观的12号原汇丰银行、13号海关大楼和20号原沙逊大厦等建筑,堪称当时上海最大也是最重要的建筑设计机构。

2013新年外滩4D灯光秀,寒风中数万人屏息凝视奇幻灯光烘托的12号和13号大楼,却鲜有人知道两幢都是威尔逊的杰作。如果没有他,如果没有外滩3号这场漂亮的上海滩首秀,外滩让人心醉的天际线就不是今天的模样。

二战后公和洋行撤离上海,直到20世纪90年代才又重返申城,凭借当年的口碑拿到包括来福士广场在内的许多商业地产项目。看来这家历史悠久的公司总能准确把握上海城市发展的脉搏。在过去这一百年间,他们两次决定从香港到上海寻找机会,都不早不晚,刚刚好。

参观指南

建议到露台餐厅喝一杯,观赏角亭和风向标,并向北眺望外滩天际线上最著名的三座楼——浦发银行、海关大楼和平饭店,都是外滩3号设计师公和洋行威尔逊的作品。

More than one hundred years ago, a young architect was assigned from Hong Kong to Shanghai to expand business for his agency. With talent and luck, he got a project, executed it superbly and thus won a galaxy of notable projects, most of which are still glistening on the crescent Bund.

The architect was George Leopold Wilson, then 32, from Messrs Palmer & Turner Co. And the first project that brought him good fortune was the Union Building, widely known today as "Three on the Bund".

In 1916 the *Far Eastern Review* newspaper called the six-story building one of the most noteworthy advances in construction and design in Shanghai - it used a steel skeleton, a technological advance that would make skyscrapers possible.

"The lines of the building are beautiful and the structure is graceful in the extreme, forming a noteworthy monument to the skill of the architects, Messrs Palmer and Turner," it said.

With a corner tower topped by golden argosy weather vane 150 feet (45.72m) above the ground, the edifice is built with skeletal steel framework, a bold new technique which was used in Shanghai for the first time.

The 46-meter-high edifice was also the tallest building in the city of then. The previous record was held by the 40-meter-high, Song Dynasty (960–1279) Longhua Pagoda for more than 1,000 years. Before the Union Building, the tallest structure on the Bund was the former Palace Hotel.

Three on the Bund was the first Bund building to be renovated for commercial use - a hub of high-end restaurants, luxury shops and art gallery. Like the Xintiandi project, the renovation that began in 2001 was controversial because it dramatically altered the original interior. Tourists and merchants loved it while preservationists disliked it and believed it opened a kind of Pandora's box bringing "mercantilism" back to the Bund.

Dating back to the 1910s, the ground, first, second, third and fourth floors were leased out as offices.

On the fifth floor were residential flats of the highest class, with roof gardens on the flat concrete roof. These flats command a magnificent view of the river and of Shanghai. The height ensured coolness and quietness, a treasure to enjoy in a city with hot summers and without air-conditioners. Renowned American architect Henry Murphy, who designed Yenching University (today's Peking University), had set up his Shanghai office on the top floor in the 1920s.

The building is also known for one of its former tenants, the Mercantile Bank of India, whose Chinese name You Li was used as the building's name.

In 1941 Japanese invaders occupied the building. The bank restarted business four years later but retreated from China in 1949. In 1953 another legendary architect moved to work in the building, Chen Zhi, director of the Civil Design Institute of Shanghai. A graduate of the University of Pennsylvania from the United States, the man was among the first to record and preserve the city's historical buildings after the 1950s.

The plinth of the building is of granite, the remainder of the filling walls being of brickwork finished with an artificial stone facing. The principal entrance was flanked by massive granite columns with carved pediments, leading to an imposing entrance hall. It is really the first masterpiece of Palmer & Turner in Shanghai.

Founded by William Salway in 1868 in Hong Kong, Palmer & Turner had a Chinese name "Ba Ma Dan Na," a Cantonese translation of its English name. As soon as the Union Building was completed, the firm moved its Shanghai office in, created a nice Mandarin name "Kung-woo," or "Fair and Harmonious" and developed dramatically in this "Paris on the Wangpoo (Huangpu) River."

Within decades the company created more than 30 excellent structures in Shanghai, including nine of the 23 waterfront Bund buildings, such as the HSBC Building and the Sassoon House. After 60 years' absence from the city since World War II, the company made its way back to Shanghai, having designed new skyscrapers since the 1990s including the Raffles City Shanghai.

In the 1920s and 30s, Palmer & Turner grew to be the largest and most important architecture design firm in Shanghai, whose works were almost a mini collection of the Shanghai architectural scene. It seems that the history-rich firm picked the right time to open a Shanghai office twice within 100 years.

Tips

Visitors can see each commercial floor. I suggest going onto the terrace for a drink where one can look up to see the stunning corner tower or overlook a perfect skyline dotted with three famous roofs of the Bund - the former HSBC Building, the Customs House and the Sassoon House, all of which were designed by Palmer & Turner.

外滩 5 号 No. 5

开启黄金时代
1920s Grandeur Starts at No. 5

1998年寒冷的一月，澳洲女人嘉娜特走进外滩5号。

"我步行到顶楼，打开一扇小门便是一个露台。哦，我的天哪！1998年的上海密密麻麻都是里弄房子，很少有人能站到这么高的地方，看到整条黄浦江这么壮阔的景观。我意识到这是个绝佳的位置和绝好机会，一定要做点有趣的事。"

一年后，她的米氏西餐厅在此开张。这家外滩当时唯一的私营西餐厅撬动了外滩的商业复苏，而1921年建成的5号大楼也曾经开启外滩大建设的黄金时代。

灰色的大楼庄严凝重，花岗岩立面中规中矩，局部点缀有雕饰的山花与希腊爱奥尼方壁柱。首位业主日清汽船株式会社主要经营长江航运业务，1907年到上海后，资本在10年间翻倍，便在外滩买地造楼。

1945年后大楼作为敌产被没收，由轮船招商局使用。1949年后由海运局等单位合用，华夏银行也曾在此设营业厅。

如今外滩5号因米氏西餐厅而闻名，但17年前外滩尚未商业开发，在当时"黯淡"的中山东一路经营西餐厅需要极大的勇气。外国朋友曾警告嘉娜特，她的餐厅如果距离由波特曼、希尔顿和花园饭店组成的三角区需要步行5分钟以上，就必败无疑。他们甚至认为有深度的人绝不会到外滩吃晚餐。不过，事实证明这个澳洲女人的直觉没有错，米氏西餐厅后来生意兴隆。

紧接着，外滩3号、18号和6号也陆续加入商业改造，外滩的夜晚不再黯淡。这段时间里上海城市发展迅速，房地产业兴盛。

马克·吐温说："历史不会重复，但有相似的韵脚。"外滩5号初建时的情形也很相似。一战后，民族工商业和金融业发展壮大，上海经济发展的引擎不断加速，房地产业空前繁荣。这段持续到二战前的岁月被称为"黄金时代"，外滩著名滨水建筑大都建于这黄金的20世纪20年代。

在群星闪耀的黄浦江畔，灰扑扑的外滩5号名气不算大，但它是"黄金时代"建成的首座大楼。自它开启新一轮轰轰烈烈的建设后，近代上海的标志性建筑纷纷崛起，而今日所见之优美起伏的外滩天际线也渐渐成形。

参观指南

米氏西餐厅的露台风景绝佳。最喜欢天空清朗、有风的日子，登临这个露台看外滩洋楼上的红旗猎猎招展。

No. 5 on the Bund, the former NKK Building constructed in 1921, was the first Bund building born in the decade of great prosperity - the "golden" 1920s.

It was built by Japanese shipping company Nisshin Kisen Kaisha, which opened a Shanghai branch in 1907 to handle business on the Yangtze River and some inland waterways. After opening its Shanghai branch in 1907, NKK's capital doubled within 10 years, enabling it in 1918 to purchase the land on which No. 5 stands.

NKK stopped business when China's War of Resistance Against Japanese Aggression (1937–1945) broke out. China Merchants Steam Navigation Co took over the building in 1945, which was later used by the Shanghai Shipping Management Bureau.

The six-story remises with a simply cut facade of gray granite stands at the corner of the Bund and Guangdong Road. The top floor houses M on the Bund, the Bund's first independently operated Western restaurant in contemporary times.

"There were lovely wooden floors and molded ceilings, still fantastic," recalls Australian restaurateur Michelle Garnaut, owner of M on the Bund, on her first visit to the building in the freezing January of 1998.

"We had to walk up seven flights of stairs because there was no lift. I opened a tiny door and there was the terrace. Oh my god! In 1998 the lane ways were all small and compact and very few people in the city could stand on the top floor to look outside at the whole river, to get such a sort of vista. It was a remarkable position and remarkable opportunity to be doing something interesting."

Despite the warnings that "no sophisticated people go to the Bund for a dinner" or "one could only open a restaurant no more than five minutes' from the triangle of the Portman, Hilton and Garden hotels," she opened M on the Bund on the roof. It was a huge success.

The 1,280 square-meter structure is designed in eclectic style by Lester, Johnson, and Morriss, one of the city's most famed architectural firms in early last century.

The firm was co-founded by legendary British architect Henry Lester who followed a doctor's advice and embarked on a journey to Shanghai after an unknown disease took the lives of his three brothers.

After working on the Municipal Council and for a real estate company, Lester founded the firm in 1913 with architect G. A. Johnson which was later joined by G. Morriss.

The value of the land Lester had purchased at a low price in Shanghai in the late 19th century soared dramatically in the building boom of the 1920s.

F.L Hawks Pott recorded the increased cost of living in Shanghai at that time in *A Short History of Shanghai* published in 1928.

"The residents of Shanghai began to feel the increased cost of living, and the dearer price of food caused by the general unsettled conditions of the world. The housing problem became quite acute, especially for those living on moderate salaries. This was due to high rents ..."

But unlike extravagant rich men of his time, Lester lived in an ordinary house and took the bus. He never married and died in 1926, leaving no heirs but a large sum of assets, most of which he bequeathed to philanthropic causes in Shanghai.

Among the string of architectural gems on the Bund, No. 5 may not be the most dazzling, but it kicked off a new decade, the "golden" 1920s during which important architecture arose and the Bund that we see today began to take shape.

Tips

Have a drink on the beautiful balcony of M on the Bund to appreciate the whole river and the famous 1920s buildings at the northern end. Pick a windy day when the skies are clear and red flags are fluttering.

外滩 6 号 No.6
通商银行的哥特式摇篮
Cradle of Chinese Modern Bank

外滩6号的小尖顶和老虎窗在外滩独树一帜，仿佛整个儿来自一个19世纪的童话。今日外滩中资银行星罗棋布，而中国人创办的第一家商业银行，就在这座童话般的房子里。

精致的小楼仅有4层，最初是美商旗昌洋行的产业，后为官办的轮船招商局购得。1897年5月27日，近代中国第一家银行——中国通商银行在这里开张，由此翻开了中国金融史的新一页，虽然此时距离首家外国银行登陆上海已有半个世纪之遥。

创办银行的晚清洋务大臣盛宣怀是个聪明人。他精选八位极有实力的华人组成董事会，又确定"用人办事以汇丰为准"的方针，拷贝不走样，照搬了英资汇丰银行全套规章管理制度。他甚至还从汇丰挖来英人担任洋大班。银行后来持续经营52年，历经晚清、北洋和国民政府的政权更迭。

银行开业的前一年，英国班尼斯特·弗莱彻教授出版了《比较建筑史》，1961年更名《弗莱彻建筑史》。这本建筑宝典囊括了世界上主要建筑的介绍，被誉为最权威的建筑通史，厚达2 000页的第20版书中提到15处上海近代建筑，6号小楼便是其中之一。书中的介绍图文并茂："其外貌像是一座晚期维多利亚式的小型市政厅，具有威尼斯自由风格又略带苏格兰男爵豪华情调。"

砖木结构的小楼原为清水砖墙，后加了粉刷饰面，看上去好像薄凉的冰糕。英国玛礼逊洋行的设计呈现维多利亚哥特式建筑风格，主入口有漂亮的连拱廊。底层窗是半圆券，二层是弧形券，三层换成平券，而老虎窗又为哥特式尖券。和谐中充满微妙变化，让古典的小楼灵动了许多。

小楼历经4年修复完成，于2006年营业，融奢侈品旗舰店和高档餐饮功能于一身，成为继外滩3号和18号后第三座时尚地标。室内幽暗奢华，但历史遗存不多。

《弗莱彻建筑史》中提到的6处外滩建筑如今仅有3处保持原貌，分别是33号原英领馆，6号和15号原华俄道胜银行。

而盛宣怀自己的巨柱门廊别墅和上司李鸿章的罗马风豪宅也入选了宝典。虽然关于上海近代建筑的介绍在这部恢宏巨著中不过寥寥数页，却浓缩了中国探索新式银行一段风起云涌的历史。

参观指南

楼上餐厅的视野不错。建议了解晚清洋务大臣盛宣怀的故事，联系外滩9号招商局小楼一起欣赏，那也是由他创立的官办企业。

No. 6 is the largest 19th century building on the bund and the cradle of Chinese modern bank.

The four-story pale gray building looks like a Gothic-style mansion with a Romanesque colonnade at the entrance. It was built as the headquarters for the powerful American traders Russell & Co, one of the earliest and most famous "Shanghailanders" on the Bund, trading in opium, tea, silk, and porcelain, etc.

In books and archives No. 6 is widely called the China Commercial Bank Building because the first Chinese modern bank was established there in May 1897. In English it was originally called the Imperial Bank of China and was renamed in 1913.

Shanghai became China's financial center in the 1870s. The city controlled 80 percent of the country's foreign trade money. In addition to dominant British banks, other countries were busy launching their Bund banking branches in the 1890s.

The China Commercial Bank (Imperial Bank of China) was founded by legendary Qing Dynasty (1644–1911) minister Sheng Xuanhuai, who famously advocated use of Western technology to save the country from destitution. Sheng was also famous as the founder of the China Merchants Steam Navigation Co and the Imperial Telegraphy Administration, also on the Bund.

Sheng was such a wise man that he modeled the China Commercial Bank on the Western banking system, especially that of the Hong Kong and Shanghai Bank Corp. He adapted the HSBC system and even hired Englishman A.M. Maitland, a former HSBC employee, as manager.

Founded with capital of five million taels of silver, the bank received deposits from local financial enterprises and the private savings of wealthy Chinese officials. It extended loans to Chinese factories such as Hua Sing Flour Factory and Long Zhang Paper Making Factory.

Starting in 1898 the bank issued bills that were printed in London in English and Chinese, bearing dragon patterns and the signature of the British manager.

After the Revolution of 1911 ended the Qing Dynasty, the Imperial Bank of China was renamed the Commercial Bank of China. After the 1920s the bank was taken over by Shanghai gangster leader Du Yuesheng supported by Kuomintang leader Chiang Kai-shek.

In 2006, the Gothic-style building that once housed China's first-attempt at modern finance was renovated into a sleek venue for fine dining and luxury shopping. The original interior was changed and no traces of the past remained.

The original red-brick walls and some Gothic architraves had all been covered

by mint-hued coatings during a four-year renovation.

However today we can still see that the designer had cudgeled brains to create a variety of different Gothic windows for different floors and the dormers, ranging from semi-circular arches, segmental arches, diminished arches and pointed arches.

This designer, with a fertile imagination, no doubt didn't know the 19th century building would survive until the 21st century, having witnessed so many Bund stories inside and in front of the rainbow of Gothic arch windows.

Tips

The restaurants have great views. It's also interesting to read about legendary modernizer Sheng Xuanhuai and visit the neighboring building No. 9, China Merchants Steam Navigation, also founded by Sheng.

外滩 7 号 No. 7
巴洛克穹顶下的电报大楼
Telegraph Building with Baroque Domes

在论及电讯引进中国的书中,外滩 7 号的出镜率非常高,因为大楼原业主丹麦大北电报公司是最早在中国经营电报业务的公司。

"第一口螃蟹不易吃。"无论电灯还是铁路,中国人对于洋玩意儿,一开始都视为邪物,担心惊扰祖先坏了风水,所以难以接受。

二次鸦片战争后,西方列强认为硕大的中国"牡蛎"已被利剑撬开,是取"珍珠"的时机了。各国纷纷要求在沿海港口铺设电线,却遭到清政府的拒绝和阻挠。

1869 年成立的大北公司无视清庭与英使谈妥只许铺设海底电缆的协议,将水线引入上海租界报房。公司于 1871 年 4 月 18 日开业,收发沪港两地公众电报,1882 年又开通了电话业务。

在靠驿站传递公文的古老中国,电报和电话的出现让"万里重洋瞬息通",带来翻天覆地的变化。北京科举考试发榜信息用马匹传递的话,至上海需要7~10天。天津、上海开通电报后，从北京到天津段采用马匹传递消息的方法，再发电报到上海，总共只需2天。

电报加快了订货速度，减少了货物囤积，并降低了买方风险。大洋行渐渐失去垄断地位，辛酸的大班们风光不再，个体经营者却因灵通的信息而发达。

大北公司生意日益兴隆，于1882年购入外滩原旗昌洋行办公楼，1906年斥资建造了这座法国文艺复兴风格的大楼。建筑高4层，造型典雅，装饰有大量的希腊爱奥尼石柱。大楼的亮点是一对法国巴洛克式的黑色穹顶，优雅地坐于顶端两侧，与白色的大山花形成鲜明对比。

1922年公司迁往爱多亚路（今延安东路）并另建大楼——如今的上海电信博物馆。而外滩7号的主人也先后换成原在外滩6号的中国通商银行和长江航运公司。1995年外滩恢复金融街功能，泰国盘谷银行通过置换入驻。

如今外滩7号的视觉焦点是门楣上一个深红色的大鹏鸟标志。这枚人身鸟翅的标志源于印度教主神之一毗

湿奴的坐骑迦楼罗，也是泰国国徽的形象。大鹏鸟头顶金色宝塔，佩戴金饰，两臂弯向头顶的姿势充满戏剧张力。泰王室常将此标志作为极高的荣誉颁赠给杰出公司。

由华人创办的盘谷银行因对泰国经济的贡献，于1967年获赠了这枚鲜艳夺目的大鹏鸟。在入驻外滩的首批外资金融机构中，盘谷银行是唯一置换了整座临江大楼的外资银行。这种敢于"吃第一口螃蟹"的精神，和最早让中国"万水千山一线牵"的大北电报公司，有异曲同工之妙。

参观指南

外滩7号仅对盘谷银行的客户开放。建议参观延安东路34号的电信博物馆，是大北电报公司卖掉7号之后建造的办公楼。博物馆免费开放，开放时间为周六、周日（上午9:30—12:30，下午13:00—16:30）。

清代末年电话交换所的电话接线生　　Chinese operators in the late Qing Dynesty

Building No. 7 on the Bund tells the uneasy story of when the telegraph was introduced to China more than a century ago.

Elevation of No. 7 is in a classic three-section both horizontally and vertically. Black domes contrast with white gables, both very Baroque.

The building was built in 1906 on the former land of American trade company Russell & Co as an office for Great Northern Telegraph Company, a Danish business, Shanghai's first provider of telegraphs and telephones.

After Shanghai opened a port in 1843, the former peaceful ancient town inevitably became a hot spot for Eastern and Western cultures to mix.

The Chinese often opposed foreign gadgets and technology at first, but slowly and cautiously tested such things as rickshaws and railways.

In 1870, when Great Northern laid a cable between Shanghai and Hong Kong, the cable at the Shanghai end was not to be landed on shore but on vessels anchored outside the limits according to agreement. But the cable at Woosung was brought ashore secretly, which was discovered by Chinese authorities and greatly protested.

However difficult the process was, the telegraphic links between Shanghai, Hong Kong and London in 1870 transformed the city's commercial import-export operations, making it possible to speed up orders, reduce stocks and diminish the risks run by buyers.

The Chinese administration initially opposed the idea but soon became convinced of its usefulness and since 1881 proceeded to install lines on their own between Shanghai and Tianjin, between Shanghai and Canton and between Zhejiang and Hankou.

Qing Dynasty (1644–1911) minister Sheng Xuanhuai cleverly employed technicians from Great Northern to purchase materials and lay the cables. A network around China was built within a dozen years.

The telegraph greatly enhanced the speed of communication in China. Formerly it took seven to 10 days to carry the results of imperial examination from Beijing to Shanghai by horses. After the Tianjin-Shanghai line was installed, it only took two days. The results were taken from Tianjin to Beijing by horse, then sent to Shanghai via telegraph.

Telegraphic communications also turned the structures of foreign firms in Shanghai upside down. The major hongs (trade companies) lost their monopoly, and the taipans were eclipsed by merchants who operated on a far smaller scale, on their own and with capital that they topped up with local loans from foreign banks.

No. 7 was designed by Atkinson & Dallas, a firm famous for creating classic European styles. A careful individual can still find its name here and there on name boards of historical buildings around the Bund.

The firm's Arthur Dallas had served in the Municipal Council and was once vice chairman of the China Architects Society and member of the British Royal Arts Society. As one of the most prolific architectural design institutions in Shanghai in the 1920s, Atkinson & Dallas had designed noteworthy works including No. 29 on the Bund. No. 7 was their first Western classic piece in the city.

As business expanded, Great Northern moved to a new mansion at No. 4 Avenue Edward VII (now Yan'an Rd E.) in 1921 and sold No. 7 to Commercial Bank of China, which was previously at No. 6 and they moved in the following year.

After World War II, buildings No. 6, No. 7 and No. 9 on the Bund all belonged to China Merchants' Steam Navigation Co, the official shipping company initially founded by the imperial government.

Yangtze River Shipping Co. occupied No. 7 after 1949, which managed a restaurant and a hospital inside the grey building. Since 1995, Bangkok Bank moved in and put a statue of a mythical Garuda on top of the entrance, which is still there today.

Tips

The building is only open to clients of the bank. However, I would suggest a visit to the Telegraph Museum at 34 Yan'an Rd E., located in the building where Great Northern relocated after selling No. 7 in 1921. The museum is free. It is open from 9:30 am to noon and 1pm to 4:30 pm on Saturdays and Sundays.

外滩 9 号 No. 9

红与灰 Red or Gray

红色还是灰色？2001年外滩9号大修启动时，专家们曾就如何修复外墙展开激烈的讨论。两派专家各持己见。红色派主张"修旧如初"，恢复小楼原始的清水红砖墙；灰色派则建议"修旧如旧"，维持已覆盖水泥砂浆的墙面，与外滩灰色建筑群保持一致。

外滩9号建于1901年，业主是中国首家轮船公司——轮船招商局。晚清洋务派于1873年创办的招商局凭借运送朝廷漕粮的优势，加上华商的支持和低廉的运营成本，顶住了外国航运公司的压价竞争。1876年招商局收购了竞争对手美商旗昌轮船公司，《申报》发表评论："从此中国涉江浮海之火船，半皆招商局旗帜。"9号小楼便建在旗昌洋行北楼前的花园里。

这也是外滩唯一现存的维多利亚风格外廊式建筑，由玛礼逊洋行设计。这种建筑风格在湿热的南亚曾十分流行。在寸土寸金的外滩，仅有三层的小楼显得颇为奢侈。

小楼初建时采用砖木结构，清水红砖墙。二、三层有美丽的敞廊，点缀着塔斯干柱式和科林斯柱式。顶部山墙雕刻精美的大山花。最奇异之处是其结构仅由8根并不粗壮的实心钢柱支撑，虽不符合现代建筑规范，却支撑了外滩9号整整一百多年。

自建成之日起，招商局一直在小楼办公，直到1937年因日军侵华而迁往香港。20世纪40年代招商局曾计划在外滩7、8、9号旧址上建造上海最高的大楼，后因内战爆发而未实现。而国际饭店设计师邬达克的档案中也存有一幢外滩40层巨厦的图纸，很可能正是该楼的蓝图。

1937年后多家公司入驻过外滩9号，新中国成立后又几易其主。到1998年小楼回归招商局时，大山花已被拆除，敞廊为玻璃钢窗封死，红砖墙也由水泥砂浆抹平，看上去俨然一座简陋的厂房。负责修缮工程的同济大学建筑系常青教授认为，当时的小楼堪称外滩最难看的建筑，"甚至没有挂上上海优秀历史建筑的铭牌。"

关于红与灰的激辩最终由西方建筑史权威罗小未先生一锤定音。她认为万国建筑应各有风采，不能千篇一律，恢复红砖外立面为好。效果图也显示红色小楼并未破坏外滩整体风貌，专家们终于统一了思想。

于是，修缮团队凿除了覆盖红砖表面的水泥砂浆，又应用德国雷马士技术进行修复，百年前动人的红砖墙终于洗净铅华。由于缺乏原始图纸，三角形大山花根据放大到极致的历史照片复制，同时参考了英国新古典主义时期以椭圆涡卷为母题的山花式样。9号小楼在外滩独树一帜的柱廊空间也修旧如初。

2012年外滩9号修复工程参加了米兰建筑展。夜色中灯光点亮小楼红砖的照片登上了意大利报纸，用璀璨夺目来形容这景象并不为过。漫步今日外滩，一排深深浅浅的灰色中骄傲地耸立着红色典雅的9号小楼，平添一丝新鲜的活力。看来这红与灰的选择，做对了！

参观指南

小楼不对外开放，但可以参观与小楼相邻的福州路17号旗昌洋行老楼，该楼建于1869年，是外滩现存历史最悠久的建筑，楼上有带露台的咖啡馆。9号小楼当年就建在洋行的花园里。

Red or gray? A debate arose in 2001 over the wall color before renovating No. 9 on the Bund, the China Merchants Steam Navigation Company Building.

This delicate villa is the only surviving Neo-Classical veranda architecture of the late Victorian era. The building showcased a colonial style of the 19th century that was once popular in hot, humid South Asian colonies.

Covering an area of 1,460 square meters, the main structure is made of brick, stone, timber and steel and supported by eight steel columns. The second and third floors feature beautiful balconies graced by Tuscan and Corinthian orders, on top of which sit giant carved gables.

However, when Tongji University professor Chang Qing took over the renovation project in 2001, "it was almost the ugliest building on the Bund and was not even listed as a 'Shanghai Excellent Historical Building'."

All the gables and sloping roof sections on the eastern elevation were dismantled. The Neo-Classical balconies were sealed by metal-framed glass windows and the original red bricks had been covered by cement mortar. Juxtaposed between the exquisite former Great Northern Telegraph Co building at No. 7 and the gorgeous former HSBC Building at No. 12, No. 9 resembled a shabby factory office.

According to F. L. Hawks Pott's *A Short History of Shanghai* published in 1928, US-based Russell Co founded the Shanghai Steam Navigation Company for shipping on the Yangtze River in 1867.

"The Chinese merchants saw the importance of Shanghai as a port and in order to secure a large share of its shipping, the China Merchants Steam Navigation Company was founded in 1872 under the initiative of Li Hung-chang. The old P&O steamer Aden was purchased, and for the first time the Chinese flag was flown over a merchant steamer. In 1877, the fleet and property of the Shanghai Steam Navigation Company was purchased by the China Merchants," Pott wrote.

In the garden that formerly belonged to the neighboring Russell & Co building, China Merchants Steam Navigation Company built the red-brick building. The Russell & Co building was built in 1857 and is the only first-generation (1840s–1850s) building to survive on the Bund.

As the principal architect of the Foreign Affairs Movement, imperial official Li was lucky enough to rely on an excellent administrator, Sheng Xuanhuai, who became his principal collaborator on economic matters.

Sheng was involved as a promoter, shareholder, or manager in nearly all the official business ventures set up in Shanghai from 1871 to 1895, including the Imperial Bank of China at No. 6 on the Bund. He became the head of China Merchants in 1873.

China Merchants kept its headquarters at the No. 9 building until it moved to Hong Kong in 1937 after Japan invaded China. Some shipping companies moved in from 1937 to 1949, after which the building was used by several state-owned organizations. In 1998, it finally returned to China Merchants as its Shanghai office, who decided to renovate the building.

Without original drawings or documents left, Prof. Chang could only study archive photos carefully to revive the original look.

This sparked the debate over the wall color. During meetings some experts suggested "obeying the old look," which was to maintain the already gray color to keep it uniform with the other Bund buildings. Other experts supported "obeying the original look," which was to restore the red-brick wall of when it was built.

At last, a drawing presenting the final effect with red walls ended the debate. The drawing showed the harmony of the Bund was not destroyed, but actually improved with red walls. Famous architectural historian Luo Xiaowei concluded that Bund buildings do not need to have a uniform color, as their beauty lied in variety. She was right.

Once that decision was made, the renovation project continued.

To recreate the original gables, Chang's team studied cartouche patterns during the British Neo-Classical period and made models to compare the depth, lines and texture of the patterns again and again.

The most difficult part was to restore the original red-brick wall that had been covered

1908年，招商局大楼外景
Merchants Steam Navigation Company Building, 1908

with grey cement mortar. If the mortar was removed improperly, the building would look like a man's face dotted by smallpox.

To minimize the damage, workers carefully removed the cement mortar bit by bit and tenderly ground the remaining traces. Special techniques from Germany-based Remmers Company were used to restore the red bricks.

After the "back-breaking surgery," No. 9 was restored to its original look in the autumn of 2004.

The former gray "factory office" now resembles the colonial-style house with lovely verandas and grand gables that it was designed to be. It is one of only two red-brick waterfront buildings, Palace Hotel at No. 19 is the other, now proudly standing amid the line of gray structures on the Bund.

Tips

Please also visit the former Russell & Co building at 17 Fuzhou Road. The two buildings are connected by a tiny corridor, which shows their historical relevance. There's a tiny, lovely café with balcony upstairs.

外滩 12 号 No. 12

外滩的橄榄叶
An Olive Leaf for Shanghai

1921年6月,外滩汇丰银行新大楼奠基仪式隆重举行,英文报纸《远东时报》将其形容为"一片诺亚的橄榄叶"。

根据《圣经》故事,诺亚乘坐方舟躲避大洪水后,他派出探听消息的鸽子衔回一片橄榄叶。这表明洪水已退,灾难终于结束。而当时上海的经济也正从一战后的衰退中渐渐复苏。

"大楼奠基象征着经济衰退的潮水正在退去。对于捱过了最困难时光的人们来说,未来是光明的。"报纸写道。

汇丰银行由苏格兰人萨瑟兰德于1864年创办于香港,因为他敏锐地意识到香港与大陆港口城市间的金融需求。银行后来从清廷的贷款中获利颇丰,一度将中国海关与盐税的存款权牢牢掌握。

奠基这一年,股东大会宣布250万港元的注册资本已增长到5 000万港元,汇丰成为远东第一银行。建造新大楼意在显示实力,重塑信心,因为金钱常随信心滚滚而来。由此银行决定,"无论要经历多少困难、痛苦或是麻烦,只要有适合大楼使用的材料,都必须想办法弄到。"银行大厅中8根由整块意大利大理石制作的奢华圆柱,便是不惜工本弄到的建材之一。

新大楼在汇丰旧楼的基础上,加上原11号别发书店和原10号新茂洋行的地皮,最后造价高达1 000万元,超出预算数倍。显著的位置,宏大的体量加上精美的建筑设计,为12号大楼赢得"从苏伊士运河到白令海峡间最壮观建筑"的美名。

公和洋行的威尔逊在1920年初便完成了设计方案,但经过一年的推敲,他居然下决心化繁为简,好像用了一块大橡皮把立面细密的雕刻统统擦去。最后落成的12号呈现简洁明快的新希腊风格。没有多余装饰,仅依赖比例和线条,就塑造出高贵的形象,时至今日气场依然十分强大。

走进银行大厅,感觉好像走进一座恢宏的神殿。八角形门厅穹顶的马赛克镶嵌画看得让人目眩神迷。8块镶板上的图画分别代表了汇丰银行在伦敦、纽约、香港等地的8家银行,每一幅都值得仰起脖子细细欣赏。关于上海的那幅画绘有一位手持舵轮、眺望远方的睿智女神,她身后的背景就是白色的12号大楼,还有英伦风格的前海关大楼等外滩建筑。

12号不仅美如神殿,当年还是一座现代化办公楼。楼内空气一小时内循环数次,冬季空气还要经过加热,比冰冷的神殿舒适多了。

1955年上海市政府搬入12号大楼办公,直到外滩恢复金融街功能而迁出。1997年,获得该建筑使用权的浦发银行在修复过程中惊喜地发现了在20世纪50年代被粉刷遮盖的八角厅马赛克画。

在1923年的落成典礼上,威尔逊演讲中称自己是"一个管弦乐团的指挥家",指挥着包括工程师、艺术家和雕塑家的团队来完成这个伟大复杂的建筑。

"建筑师的作用就好像指挥乐团演奏一首自己的作品一样。如果演出成功了,那么要感谢每一位参与的人。因为只要有一个'错音',就足以导致一片混乱而非如此和谐的效果了。"

威尔逊90多年前指挥的这件作品不仅在上海建筑史上留下灿烂的一笔,从经济角度看也圆满完成了使命。自它奠基之后,上海的民族工商业和金融业发展壮大,城市发展的引擎加速,从而迎来一个生气勃勃的"黄金时代"。也许这栋壮美的楼,真如一片飘落至外滩的"橄榄叶",为这座一度低迷的城,带来信心与希望。

参观指南

浦发银行一楼在营业时间对外开放。

No. 12 on the Bund, the former HSBC Building now housing the Shanghai Pudong Development Bank, was described at the time as "an olive leaf" carried by Noah's dove, signifying a bright future for the great business houses or "hongs" in China.

"The laying of a corner stone for the magnificent new home of Way-foong (HSBC) ... is the olive leaf that symbolizes for the 'hongs' of China that the waters of business depression are subsiding and that the future holds bright for those who have passed through the flood of hard times," reported the *Far Eastern Review* in June 1921.

The Hong Kong and Shanghai Banking Corporation Building was completed in 1923 and known as the most luxurious building between the Suez Canal and the Bering Strait. It has a floor area of 62,000 square feet (23,415 square meters) and was said to be the second-largest building in the world at that time, after the Bank of Scotland building in the United Kingdom.

The white, Neoclassical edifice (a seven-story central section with five-story sections on each side) dominated the Bund as its tallest building, until it was surpassed by Sassoon House and the Customs House erected in the late 1920s. It was also the most massive and the most magnificent, on a prime and auspicious location. The facade was covered with white Hong Kong granite.

It was built on the site of the old HSBC house at No. 12, the book store Kelly & Walsh at No. 11 and Messrs Thomas Simmon & Co at No. 10, a trading company.

Rising more than 50 meters to the massive dome, the building stood out in clear view of the merchant ships from around the world sailing up and down the Huangpu River.

George Wilson, chief architect of Palmer & Turner, designed the building in a Neo-Grec style, without ornamental carving or sculpture in a preliminary plan. Relying almost entirely on proportion and lines, the building achieved great dignity and beauty with simplicity. The initial plans were far more ornate and less appealing.

The Hong Kong and Shanghai Banking Corp was established in March 1865 in Hong Kong and in Shanghai a month later. According to HSBC archives, the inspiration behind the bank's founding was Thomas Sutherland, a Scot then working for the Peninsular and Oriental Steam Navigation Co. He realized the considerable demand for local banking facilities in Hong Kong and on the mainland coast.

HSBC had lent a large sum of money to the imperial Chinese government and in this way it took control of the salt tax and tariff in China. The bank's Chinese comprador was even awarded a royal official cap button as an honor given by the Qing (1644–1911) royal government.

In 1923, the *Far Eastern Review* reported that the bank's original US$2.5 million capital was raised to US$50 million at a shareholders' meeting in May 1921.

In 60 years, the little treaty port commercial bank had expanded into the foremost financial institution in Asia, the strongest foreign bank in the British Empire.

20 世纪 20 年代 汇丰银行附近街景 The Bund near the HSBC Building in the 1920s

That explains the grandeur of the building. First and foremost, the magnificence of the entrance hall and the ceiling mosaics is awe-inspiring, and visitors look up in admiration for so long that their necks get sore.

The 15-meter-wide octagonal ceiling suggests a sacred temple, inspired by the Chinese belief that the number "eight" would bring good fortune and prosperity. (The word for eight, ba, sounds similar to the word for prosperity.)

The ceiling is supported by eight detached Sienna marble columns. The outer arcade is of the same marble, but the bases and capitals of the columns are of bronze.

The most exciting part is the domed ceiling decorated with rich Venetian mosaic.

The center depicts the mythological figures of Helios, Artemis and Ceres. The eight principal panels represent the banking centers of the East and the West, including Shanghai.

It is said former Shanghai Vice Mayor Pan Hannian made the decision to cover the mosaic with a thin coating of white painted plaster during the 1950s renovation that turned the building into the municipal government headquarters.

Thanks to Pan, it was sensational news when the plaster was removed in 1997 to reveal the splendid mosaic, reviving memories of the Bund in its glory.

The octagonal entrance hall opens onto the main banking hall, a vast space of perfect proportion and light. Four monolithic columns, two at each end, were worked in Italy and delivered to the bank without damage. Each column weighs around seven tons.

It was reported by *Far Eastern Review*, "No difficulties, no pains or trouble have been spared in finding and securing whatever seemed most suitable for different parts of the structure. The crafts and trades of all nations have been employed."

The bank was constructed with the latest technology and amenities, so it was as grand as a European palace, but much more comfortable.

It had the most up-to-date system of ventilation and warming; fresh air was drawn in at certain points and washed by passing through a water stream.

In winter, this cleansed air was warmed and pumped through a system of ducts into the rooms. The air in the building was recycled twice an hour in the winter and six times an hour during the summer.

At the opening ceremony in 1923, chief architect Wilson described his role in designing the great modern building as that of "the conductor of an orchestra" who "gathers about him assistant architects, engineers, artists, sculptors and other specialists, carefully selecting those he knows will work in sympathy with his ideas and aspirations."

"It is for the community to decide whether the architect has justified the

20世纪30年代，汇丰银行门前的铜狮引路人驻足观看
Street View fronting the HSBC Building in the 1930s

confidence placed in him and produced a building worthy of the opportunity which, as regards the size, site and in other ways, was an exceptional one. It is sufficient for him to say that he has given of his best."

After 1949, HSBC moved to 185 Yuanmingyuan Road. The building was renovated in 1955 to house the Shanghai Municipal Government.

In 1997 it was renovated again, based on the original blueprint, to house the Shanghai Pudong Development Bank.

And so, the dominant white-domed building fulfilled its mission, bringing hope to the business community of Shanghai for at least a decade. In the 1920s and early 1930s, the sunshine was breaking through the gloom and commerce flourished again after the "olive leaf" on the Bund was planted.

Tips

The building is open in office hours as the Bund branch of PDB Bank.

外滩 13 号 No.13

外滩钟楼之巅
Bell Tower atop the Bund

外滩的浪漫,在徐徐江风轻掠过华丽的洋楼。而海关大楼整点敲响的钟声也为此情此景增色不少。

1927 年,海关大楼建成,成为当时外滩最高的建筑。

海关大楼建筑风格的演变浓缩了一段上海城市发展史。1857 年,江海北关正式迁入的中式歇山顶建筑酷似电影里青天老爷办公的衙门;1891 年翻建时,中式大屋顶变为都铎式的红瓦砖墙,透着浓浓的英伦风情。

当时光流转到摩登的 20 世纪 20 年代,这两种风格都过时了。公和洋行的威尔逊选择了有节制的古典主义风格,简洁而现代,又带一点当年最时髦的装饰艺术派的味道。

新大楼高 8 层,规模庞大,造价高达 425 万两白银,花费 30 个月建成。大楼立面以花岗岩作装饰,面向外滩的主入口柱廊是纯粹的多立克式,灵感来自古希腊雅典的帕提农神庙。

在外滩万国建筑群的中心,最吸

引眼球的是12号原汇丰银行和13号海关大楼，它们的设计方案都出自建筑师威尔逊之手。建筑师显然考虑到相邻而建的视觉效果——海关大楼的垂直线条突出了高度，与12号的横线条形成有趣对比。虽一横一竖，但两座楼的构图皆走简约风，并无雕梁画栋，看起来十分和谐。外滩4D灯光秀也以这两座大楼为屏，它们既对比又和谐的关系营造了特别奇幻的效果。

据1927年《远东时报》报道，当时装饰艺术风的摩天楼已开始流行，充满动感的风格与上海城市的活力生机非常相称。而新海关大楼正是一件转型期作品，既有摩天楼般的钟楼，又保留着古典的多立克柱廊。

今日走进海关大楼，好像走进一座迷宫，这么多的房间，这么多的电梯，这么多的门。令人欣慰的是建筑细节保留完好，也许是因为外滩13号自建成后一直作为上海这座城市的海关办公地，建筑功能从未改变。

海洋主题的大厅有红褐色贝壳图案，天花板上装饰有八幅生动的彩色马赛克天顶画，描绘了船只和海浪，航海时代的气息扑面而来。原来在大厅的中央还悬挂着一支水晶吊灯，用来照亮彩画，可以想象这海关的大厅有多么璀璨。

虽然海关迷宫让人流连，但最激动人心的还是攀登钟楼。海关大钟是仿照伦敦伊丽莎白塔大报时钟设计的，并由英国公司制造。"文革"期间，原先报时的英国名曲《威斯敏斯特》改为由扩音器播放《东方红》。如今，每逢整点的钟声还是由1927年安装的这口大钟敲响的。

通往钟楼的台阶既陡又长，没有空调和通风设备。在高温闷热中攀爬近百级台阶，方才到达这外滩钟楼之巅。

凉风倏然袭来，深呼吸中一览外滩众楼小，身世不凡的洋楼们都化作一片低矮的灰色背景。到达这个高度后，似乎终于从外滩的喧哗和物欲中抽离，内心格外澄净，也许这就是人生需要登高望远的原因。迷思中，身旁英国大钟轰然敲响，这声音与在地面上听到的很不一样，震撼鼓膜与心灵。此刻，站在这外滩钟楼之巅，真是极致的生命体验。

参观指南

不对外开放，但可以欣赏海洋主题的大厅。

No building on the Bund tells the city's architectural history like the Customs House, which was built three times.

The first Customs House on Hankou Road was designed like a Chinese temple with a large roof and upturned eaves. It was built after the city opened its port in 1843 and was necessary to collect customs duty.

Built in 1891 on the same site, the second Customs House appeared to be a red-brick, Victorian building with a six-story clock tower as the centerpiece.

And No. 13 on the Bund, the third version completed in 1927, was treated in a restrained classical style.

The eight-story Customs House, topped by the clock tower standing 90 meters high, was designed by Palmer & Turner architects and took 30 months to construct. The entrance portico is in pure Doric style, the inspiration being taken from the Parthenon in Athens. The entire eastern section is faced with granite.

Covering one city block, the Customs House became the tallest building on the Bund when it was built.

The tower was so high that it was separated from the hustle and bustle of the Bund, overlooking the ships coming and going on the Huangpu River. It was a huge investment of 4.25 million taels of silver and was built on a grand scale with a large number of architectural details.

Vertical lines predominate from the third to the seventh floor to accentuate the height. The lines are a pleasant contrast to the long, horizontal lines of the former HSBC building at No. 12, which has a much greater frontage on the Bund. Both buildings are creations of Palmer & Turner, designer of nine waterfront Bund buildings.

According to English newspaper *Far Eastern Review* in 1927, the early Modernism deeply stamped by European influences began to give way to the vertiginous, pure lines of skyscrapers with exuberant Art Deco ornamentation after 1925.

"This dynamic style, inspired by the New World, was nicely in keeping with Shanghai's vitality. The new Customs House marked the transition. It preserved a Doric doorway, but its clock tower, a replica of London's Big Ben, rose to a height of 85 meters," it said.

The bustling, complicated building is filled with staircases and rooms and today houses more than a dozen organizations and companies, including Shanghai Customs, Shanghai Extry-Exit Inspection and the Quarantine Bureau and China Ocean Shipping Agency. It is easy to get lost and hard to get directions because even the staff are not familiar with all the rooms.

Archives show that No. 13 housed numerous officials and departments, including the chief tide surveyor's office, box holders, transport officers, carpenters' shop and car garages. There were flats for senior staff and their servants. The Whangpoo Conservatory and the Bank of China also maintained offices.

The Customs House is one of the best preserved historic buildings on the Bund. The vast entrance hall is filled with maritime patterns. The walls are paneled in various selected marbles and the piers are covered with the same material inlaid at the corners with mosaics.

In the center of the hall is an octagonal ceiling containing eight vivid, multi-colored mosaics of Chinese junks and sea waves. Originally a magnificent crystal lamp stood as the centerpiece of the hall, illuminating the mosaics on the ceiling.

However beautiful the mosaics, the most breathtaking part of exploring the Customs House was climbing a long, steep, spiral iron staircase to the top of the clock and bell tower, which contains a clock and bell replica of London's Big Ben, formerly nicknamed "Big Ching."

It was a hot, sunny day but there were no fans or air-conditioning during the journey upward which was dizzying, suffocating and a bit terrifying.

After climbing countless stairs, I finally reached the top, where a breeze from the river was blowing my hair. I was melting and my heart was throbbing, but my mind was cool and crystal-clear in the breeze at this height.

I had never seen the Bund from this position, so high and in the center. The grand buildings seemed short and small in a harmonious gray conglomeration. They were created by various architects in different styles. There was no time for the Municipal Council to enforce harmony in construction, yet the buildings in their diversity blend beautifully. The Bund is a gift.

At 11am the familiar melody *The East*

20世纪20年代的新海关大楼
The Customs House in the 1920s

Is Red (instead of *Westminster*) rang out from four speakers and then the bronze bell next to me tolled 11 times.

It was so loud that my ears and my heart were shaken. What an extreme experience! Remember, the bell toll was the same sound that marked time along the Bund for 88 years. Since then, listening to the hourly chiming of the Customs House Clock has been one of the most romantic experiences on the Bund.

Tips

The building is not open to the public, however you can visit the sea-themed entrance hall.

外滩 14 号 No. 14

最年轻的老大楼
The Youngest Old Building on the Bund

外滩临江历史建筑中,外滩14号拖到1948年10月才造好,是最年轻的一座老大楼。虽然位于中心位置,大楼却从未成为视觉的焦点。灰白水泥外墙与左右气派的花岗岩立面相比朴素至极。

14号原先的建筑是文艺复兴风格的德华银行大楼,一战后由交通银行接管。交通银行1908年由清政府邮传部创立,曾经为多条中国铁路的赎回和建设筹资。1928年为了靠近新成立的南京国民政府,交通银行将总部从北平迁往上海并入驻外滩14号,1937年委托匈牙利建筑师鸿达设计新楼。

鸿达的名作包括新新百货和国泰电影院。同为在沪打拼的东欧老乡,鸿达的职业生涯与国际饭店设计师邬达克有不少相似之处。当时的上海约有20国公民享受治外法权的保护,即外国人在华犯罪可免于中国法律的审判。可是邬达克和鸿达因为来自解体的奥匈帝国而无法享受这个待遇,所以工作起来格外小心。因为一旦发生纠纷打起官司,两人在中国法庭上肯定是要吃亏的。

每枚硬币都有两面。也正因如此,华人精英更青睐这样"中立的"建筑师,两人由此打开了华人市场并取得成功。

一战后外滩经历新一轮翻造,欧美风行的"装饰艺术风格"开始流行。鸿达也为14号设计了饰有垂直线条的"装饰艺术派"立面,外墙仅由水泥覆盖。虽然简素其外,其实精美其中。

从福州路侧门进入大楼,二层宽敞的大厅分布着近20根巨大方柱,宛如一座富有现代感的"神殿"。天花板由许多方格装点,每块方格都嵌有一块通透的玻璃。淡彩色的水磨石在地上拼成图案,与天花板和方柱相映生辉。

设计师似乎偏爱水磨石这种并非昂贵的材质。他在墙面上也大量使用极具凹凸感的黄色水磨石,与地面浑然一体,再配以黑色水磨石勾边,打造出强烈的几何效果,这也正是当年时髦的"装饰艺术派"的标志。

工程曾因抗日战争停工,后来由华盖事务所修改后建成,但交通银行并未使用多久就将总部迁回北京。自1951年起上海总工会开始在14号办公。

作为外滩历史建筑中最年轻的一员,富有现代气息的14号并不惊艳。但若结合一段时代背景细细欣赏,便不难体察这简洁而又精致的、不俗的美丽。

参观指南

不对外开放,但可以欣赏水磨石门厅的精美设计,特别是富有艺术感的楼梯。

The last historical building constructed on the Bund, No. 14 completed in 1948, is considered a typical example of Art Deco in white concrete with powerful vertical lines, a simple facade, and black marble door frames. The interior is opulent, colorful, and very Deco.

The eight-story structure used to be the China Bank of Communications Building and today houses the Shanghai Municipal Trade Union Council.

In the middle of a row of 23 exquisite buildings on the Bund, it has never been a centerpiece.

Juxtaposed with buildings with marble walls and grand columns, the simple, plain facade of No. 14 wins little attention from those who prefer more ornate structures with architectural flourishes. But it embodies Deco.

The land for No. 14 and No. 15 was first rented by British opium trader Lancelot Dent in 1844 and then taken over by Deutsch-Asiatische Bank, formed by more than a dozen German banking corporations in 1890.

In 1902 German architect Heinrich Becker renovated the facade and added a sloping roof to make it more practical and attractive.

In 1917 near the end of World War II, China's state-owned Bank of Communications took over the property.

The bank first opened an office on Tianjin Road, then operated temporarily in the Customs House at No. 13 before moving to No. 14.

No. 14 at first was only a Shanghai branch, but in 1928 the bank moved its headquarters to Shanghai from Beijing to be closer to the newly established Kuomintang government in Nanjing. The bank had financed the construction of more than 50 railways in China and was one of the first Chinese banks to expand overseas, with offices in Hanoi in Vietnam and Rangoon in Burma (now Myanmar).

The old four-story building was far too small to accommodate the headquarters of the gigantic Chinese bank.

Famous Hungarian architect C.H. Gonda designed the new Art Deco building in 1937, but construction was postponed until the end of the War of Resistance Against Japanese Aggression (1937–1945).

Allied Architects, a famous Chinese firm founded by three Chinese architects, Later revised the original design. Construction was completed in October 1948.

It's a typical Art Deco building with a lot of vertical lines. The facade is symmetrical with its top shaped like a small tower. The external walls are covered by white cement. Black marble is used extensively on the street-level facade.

The interior of No. 14 in parts is evocative of the works of famous architect L.E. Hudec, who had a Hungarian background. Hudec designed the Park Hotel and other Shanghai buildings.

The entrance hall is imposing, with many architectural details. It contains two curving staircases with artistic copper railings.

The grand hall on the second floor is even more spectacular. Nearly 20 tall, square columns give a feeling of a Greek temple with a modern touch. The flooring is patterned in yellow, green, white and pink terrazzo. The coffered ceiling is decorated with colorful glass.

Other parts of the building, both floors and walls, are lavishly adorned with yellow and black terrazzo in geometric patterns favored in Art Deco works.

The Art Deco vogue spread to Shanghai soon after it was born in France in 1925. Around a quarter of the historic buildings on the Bund contain Art Deco features.

But the Bank of Communications did not use the building for long. Shanghai Trade Union Council began using the Art Deco building from 1951 until today. It has links with workers' unions in 26 countries and regions.

No. 14 is not a building you would love at first sight, but behind its simple exterior are quite a few inspiring architectural details. It is also well preserved. Most importantly, it was the last stroke on our painting of the historical Bund. When it was completed in 1948, the Bund we see today had finally taken shape.

Tips

The building is not open to the public, but visitors can admire the entrance hall and artistic staircases.

外滩 15 号　No. 15

先锋的古典之作
The Avant-garde, Classical Building

1902年秋天,外滩15号华俄道胜银行新楼竣工,舆论称其超越了"地球这一隅迄今为止所取得的建筑艺术"。主持设计的德国建筑师倍高年仅34岁。

如今这座大楼是上海现存最早的西方古典主义建筑。19世纪的外滩清一色的砖墙建筑,还未有这种以天然石材为主要材料、立面呈经典三段式的大楼。大楼兴建过程中恰逢"义和团运动"爆发,工匠们逃离工地,以致工期拖延。老牌建筑师们便认为这个大胆创新的年轻人要吃苦头了。

2010年开始修缮15号的中国建筑师们不断地体会到这位百年前同行的前卫与革新。当年钢筋混凝土尚未在华推广,但建筑师居然想到将碎砖、石头和水泥混合使用,虽牢度不比混凝土,但效果也很棒。

在英籍建筑师希尔的大力协助下,倍高摒弃以纸筋灰浆粉刷的惯用手法,用花岗石作为外墙饰材,并在表面镶贴乳白色的釉面砖,开创了先河。

郑时龄院士认为15号的原型是迦布里尔在法国凡尔赛宫花园中为蓬巴杜夫人修建的小特里阿农宫,也是上海最早按西方古典主义章法运用柱式的实例。

继外滩15号之后,倍高的事务所设计了大量德资项目,包括后来被拆除的外滩23号德国总会。

笔者曾有幸探访修缮中的15号。黄昏的微光透过彩色玻璃天棚,轻轻点亮了宽阔的楼梯和若隐若现的浮雕。三楼301室,雕刻着众多动物图案的木墙裙席卷了每处墙面,动物形象刻画得细致灵动。隔壁300室则风格迥异,褐色木雕变为金色壁饰和巨幅油画,俄国宫廷气息扑面而来。

兴建大楼的华俄道胜银行是清政府唯一参股的合资银行,总行在俄国圣彼得堡。1928年中央银行成为15号的新主人,1949年后上海市航天局等单位曾在此办公。1994年中国外汇交易中心通过置换入主15号,之后上海黄金交易所也在此成立。

今日的外滩15号看上去纯粹而古典,一个世纪前却是跨时代的先锋之作。对此,1902年一家德文报纸这样报道:"毫无疑问,这绝对是有史以来整个中国沿岸最结实耐用且最为昂贵的建筑。德国建筑师们完成了这个任务,其难度是外行根本无法想象的。设计方案巧夺天工,施工精益求精,这为两位辛劳的建筑大师——希尔和倍高赢得了最高的荣誉。"

参观指南

不对外开放,但可以细细欣赏修复过的外立面,或在门口眺望一眼美得让人惊讶的大厅。

No. 15 on the Bund is the city's earliest surviving Western Classical building, a bold design by a young architect 113 years ago in 1902.

Tired of the plaster facades popular in Shanghai over a century ago, 34-year-old German architect Heinrich Becker used heavy marble and cream-colored enameled tiles to decorate the facade of No. 15. His approach was bold at the time and created a magnificent effect.

The four-story, 5,018 square-meter structure was built for the Russo-Chinese Bank, the only joint-venture bank between the Qing Dynasty's (1644–1911) royal government and foreign capital.

Founded in St Petersburg in 1895 with Russian and French capitals, the bank was later joined by the Chinese government as a partner to fund construction of the China Eastern Railway.

In 1899, the bank bought the site of No. 15 from the British company Dent & Co and commissioned Becker to build a new Shanghai branch. He was assisted by British architect Richard Seel.

At that time there was not a classic-style building built with stones on the Bund yet. Experienced architects had laughed at the "fancy idea" of the young architect.

Moreover, driven by the panic of "the Boxer Rebellion," Chinese construction workers ran away from work. Becker had overcome great difficulties to complete No. 15, which was the first Shanghai building that could compete with European counterparts in terms of design, materials and construction. He became the first man to introduce 100 percent European classic architecture to China.

However, the Russo-Chinese bank went bankrupt and the Central Bank of China founded by the Nanjing Kuomintang government became the new owner of the building in 1928. After 1949, No. 15 housed state-owned institutions including several democratic parties before it became the home of the China Foreign Exchange Trading Center, which was launched in April 1994.

It was the first banking organization to return to the Bund, the former Wall Street of the Far East. The Shanghai Gold Exchange also opened here in 2002. The building reopened last year after a successful renovation.

The facade is graced by two Ionic columns, four pilasters and two pair of Tuscan-order columns underneath. The architect may have been inspired by the Petite Trianon, a small, delicate palace built with white limestone and rosy-hued marble in the garden of the Versailles Palace.

The entrance hall contains numerous towering columns and a W-shaped white marble staircase leading to the third floor, which had a stained-glass ceiling. The ceilings are more than five meters high.

Two rooms on the third floor are impressive. Room 301 is embellished lavishly with chestnut-hued carved wood for the wainscotting, door frames, furniture and the entire ceiling. The vivid animal patterns on wainscoting are amazing: each animal is distinct.

Room 300 next door is in another style. Both the pilasters on the white walls and the white ceiling are highlighted with golden patterns. Like many Bund buildings, the third floor features a lovely balcony that allows visitors to appreciate the river scene in front, with a humid breeze from the "Whampoo River."

The creator of No. 15, architect Becker, had studied architecture in Munich and lived in Shanghai from 1898 to 1911.

His firm Becker & Baedeker in cooperation with Karl Baedeker designed a range of heritage buildings, including the Deutsch-Asiatische Bank at No. 14 and the Club Concordia at No. 23. Both were demolished, the first to make way for the Bank of Communications, the second for the Bank of China.

Only No. 15, on which he had lavished so much effort, still survives on the Bund.

Hard work paid off. When it was completed in October 1902, a German newspaper published a report calling Becker a dare-to-think architect:

"It was undoubtedly the most solid and expensive building on the whole Chinese coast in history. It was really a breakthrough of architectural achievement. German architects had completed the mission after battling incredible difficulties. From all aspects the building was well-designed and delicately constructed, winning the ultimate honor for two hard-working architects R. Seel and H. Becker."

Tips

The building is not open to the public, but the restored facade can be appreciated and a glance into the entrance hall will amaze you.

外滩 16 号 No. 16

外滩的"希腊神庙"
Greek Temple on the Bund

1927年的外滩分外热闹。英文报纸《远东时报》同年6月的一篇关于"外滩新一轮巨变"的报道中提及："几乎每个商业或住宅区都在兴建办公楼、公寓楼或者商店。照此速度下去，不久上海就会出现一条黄金商业街。"

文中列出的外滩新厦包括刚竣工的16号台湾银行。在这条"远东华尔街"上，16号并非最雄伟的银行建筑，但其古希腊神庙式的外观独一无二。

这座矩形建筑与古希腊供奉神像的神庙一样，由巨大的石柱环绕，气势非凡。虽然"外滩神庙"的雕饰较为简素，却有自己的特色。爱奥尼柱头两端状如"发卷"的曲线涡卷改为直线勾勒的"回"纹图案。这种富有东方色彩的"回"纹柱头在外滩12号和27号也能看到，可能是当年的一种设计潮流。

大楼原址为一座东印度式假4层砖木建筑，长期租借在此的日资台湾银行洽购成功后，于1924年拆除原楼，聘请英商德和洋行设计新楼。

台湾银行是一家日本官商合办的银行，甲午海战台湾割让给日本后，于1899年开设，1911年成立上海分行。在台湾日据时期，该行一直扮演着发行"台湾银行券"的关键角色。二战后，大楼先后由中国农民银行和上海市工艺品进出口公司使用。1998年招商银行开设外滩支行，16号重新成为一间忙碌的银行。

虽然几经改造，室内还保留着典雅的大方格石膏天花、细密的黑白马赛克和考究的巧克力色墙砖。一楼大厅傲然矗立的两排大理石方柱，让人感受到"神庙"的气场。

初建16号的1927年，今日所见之外滩万国建筑群已基本成型。《远东时报》记者写道："这些风格各异的建筑由来自不同国家的能工巧匠设计，而工部局从未要求过以某种风格来和谐统一。也许是巧合，这么多种风格建筑配在一起的效果竟然相当悦目。近年新建的横滨正金银行、台湾银行、字林西报大楼、汇丰银行、海关大楼和沙逊大厦都极好地融入其中。"

在热闹的1927年，列柱环绕的16号大楼加入了外滩建筑群，左侧是状如法国凡尔赛小特里阿农宫的华俄道胜银行，右为字林西报颀长的文艺复兴风格大楼。外滩唯一的"希腊神庙"与左邻右舍，与这条黄金天际线上其他形形色色的建筑融为一体，浑然天成。

参观指南

招商银行一楼在营业时间对外开放。

The Bund is a huge mirror reflecting the Shanghai incarnations of world-famous buildings. While No. 15 was likely inspired by the Petit Trianon in Versailles, next door, No. 16 on the Bund, resembles an ancient Greek temple with four two-story columns on the waterfront facade. The building today has four stories, with the facade in gray marble.

The site belonged to Shaw Bros & Co in the 1860s. HSBC owned the site since 1886 and added two floors to the previous two-story veranda building for lease. The tenants included the famous architectural firm Morrison & Gratton, German trade company Telge & Schroeter and the North China Insurance Co. The Bank of Taiwan had leased the building before they bought it from HSBC in the 1920s.

The Bank of Taiwan was established in 1899 as Taiwan's central bank by Japanese authorities, occuppying Taiwan at that time. The bank played an important role in issuing official currency and supporting Japanese companies during the Japanese occupation of China.

In 1911 the bank opened a Shanghai branch in the veranda building and built the grand new building in 1924 that covered 973 square meters.

In 1927 when the new building was completed, the bank published an advertisement in the English newspaper *Far Eastern Review*, stating that the Taipei-based bank operated branches and agencies across Japan, China, Java, India, London, New York, Hong Kong, and Singapore. Moreover, they had "correspondents at all the chief commercial cities of the world."

The new office was designed by Lester, Johnson, and Morriss in a "simplified Greek temple style." The building is partially enclosed by grand orders, including four columns on the facade and six pilasters in composite style on the southern side. The second and the fourth floors are topped by simple-cut architraves. Above the third floor is an overhanging cornice.

Tongji University professor Qian Zonghao discovered that the designer had used a key pattern to replace the usual spiral volute of the capitals of the building's Ionic orders. He has identified the same pattern on the capitals of columns in the Jardine, Matheson & Co Building at No. 27 (1922) and the HSBC Building at No. 12 (1923).

The three buildings were designed by different firms but built in the same period. This implies that the pattern did not reflect individual taste, but an architectural vogue at the time.

Qian adds that the interlaced vertical and horizontal lines in geometric fret patterns were used by Egyptians to decorate the ceilings of mausoleums and by Greeks to decorate pottery. Ancient Chinese also used decorative fret patterns extensively.

He assumes western architects might use the pattern as a symbol of Oriental art, which resembles the square-shaped dou gong (wooden bracket set) commonly seen on traditional Chinese buildings.

The Bank of Taiwan was taken over by

the Chinese government after Japan surrendered at the end of World War II in 1945.

According to Huangpu District archives, the building was used by the Farmers Bank of China for several years and since 1955 was used as the office of the state-owned Shanghai Crafts Import and Export Co. The China Merchants Bank took over the building as its branch on the Bund in 1998 and operates there today.

The ground floor is still a busy banking hall today. It retains the original, highly ornamented coffered ceiling. The staircase leads to the upper floors for wealthy clients, and the original mosaic flooring, as well as the chocolate-hued tiles, that adorn half the walls are well preserved.

After the building was completed in 1927, architects Lester, Johnson & Morriss moved their office into the building.

The year 1927 was also eventful for Bund reconstructions. *Far Eastern Review* published a story in June 1927 stating "The Bund is undergoing yet another quick transformation at the present moment."

It listed several newly or nearly completed buildings, including the new Customs House at No. 13, Sassoon House at No. 20, the NKK Building at No. 5 and the Bank of Taiwan at No. 16.

"Building in Shanghai, in fact, is progressing at an amazing rate, considering the

20世纪20年代的台湾银行大楼
The Bank of Taiwan Building in the 1920s

general business depression. One can scarcely travel a block in the business or residential sections of Shanghai without seeing construction work on office buildings, apartment houses or shops ... At the current rate of progress, it will not be long before Shanghai can boast a billion-dollar shop-front," the report predicted.

And history has proved the prediction.

Tips

The banking hall on the ground floor is open to the public.

外滩 17号 No.17

大力神托举的报业大楼
Press Tower Supported by Atlantes

1935年早春三月,项美丽坐游轮到上海。这位曾经闯荡非洲的美国女作家原本只打算游玩数周,不想却喜欢上"魔都",很快在外滩17号的《字林西报》找到工作,在江西路租了房子住下来。

《字林西报》曾是在华出版的最著名的英文日报,前身为周刊《北华捷报》。从1850年到1951年,这份报纸记录了上海城市历史中最跌宕起伏的一百年,曾经骄傲宣称自己是"在每栋商务楼或领事馆都能看到的报纸,同时还是媒体引用最多的报纸。无论质量还是发行量,都名列前茅"。

由于业务发展迅速,1921年报社兴建新楼。德和洋行充分利用外滩这块狭小的地皮,设计了一座身材颀长的报业大楼,于1924年落成。

外滩17号楼体苗条修长,屋顶上建造了具有巴洛克意味的双塔。不仰头细看,很难发现高悬于檐口下的8尊大力神雕塑。

设计师用意大利花岗岩作雕刻材料,但只雕塑了大力神的上半身,腰部开始自然过渡到涡卷形的托座,宛若蜷曲的大腿,真是奇思妙想的设计。这8个大力神都高举双臂托举檐口,充满了艺术的张力。他们灵活生动的姿态各不相同,强壮手臂的肌理清晰可见,劳苦功高地举了90年,未曾松懈。

当年《字林西报》编辑部位于5楼,其余楼层租赁给友邦保险等公司。1941年17号为日军所占,1951年后由中国丝绸进出口公司等单位使用。1998年友邦保险重新入驻17号,成为第一家回归昔日"远东华尔街"的外资金融机构。

爱冒险的项美丽在自传中回忆在17号上班的日子,虽然挑战不足,但相当充实。报社常安排她专访退休的英国"大班"或报道一家泳池开业。有时她也会自己选题,比如"中国药店用真正的印度支那树獭来招徕生意"。她的社交活动也丰富多彩,所以在上海的日子"满满的"。

项美丽还爱上了已婚的新月派诗人邵洵美,甚至嫁他为妾,在沪上西侨社区中引起一片惊疑。她在自传中也提到"只有一次洵美到《字林西报》编辑部来找我。他的苍白面孔和一身长衫在那些英国记者中引发一阵骚动,连他自己也感觉到了,后来只约我到外滩见面"。

这份恋情没有结果,1940年项美丽离开上海。利用在沪积累的人脉,她采访了宋霭龄和宋美龄,写出畅销书《宋氏三姐妹》,一举成名。从此,她再没有回到过这座城市。

参观指南

一楼对外开放,建议仰头欣赏托举檐口的大力神雕塑。

20世纪30年代的《字林西报》排字间　Linotype room of *North-China Daily News* in the 1930s

American writer Emily Hahn got her first Shanghai job at the *North-China Daily News* at No. 17 on the Bund after falling in love with Shanghai in 1935.

No. 17 was erected in 1924 as the editorial office for the city's first English-language daily newspaper, *North-China Daily News*. It was "the newspaper which is found in every office, consular or commercial; it is the newspaper most frequently quoted by both the foreign and vernacular press; and it is the one newspaper which combines a quality with a quantity circulation - being the largest both in size and circulation in China."

Hahn, who already had lived an unconventional life of adventure and travel, took the job at the paper, rented a flat in the nearby Kiangse Road (today's Jiangxi Road M.), and began to live a full life in swinging Shanghai.

"My days were crowded. Usually my day's assignment could be polished off in the morning. It might be an interview with some retiring magnate, or perhaps a swimming pool was being opened by an advertising club. Or I might dream up a piece myself, about a Chinese drugstore that hung cages of real Indo-Chinese sloths around to attract trade...I could write it up in the office or at home," she writes in her autobiography *China to me*.

Hahn became the mistress of Chinese

poet/publisher Sinmay Zau, who was one of the main attractions of her Shanghai stay.

"Once and only once Sinmay called for me in the North-China office: his pale face and long gown caused such excitement among the mild British reporters that he became self-conscious and after that made me meet him out on the Bund," she writes.

As the newspaper was expanding rapidly, a larger, modern office was required. Architects Lester, Johnson & Morriss designed a tall, narrow building to make full use of the only 900-square-meter plot on the Bund.

With Neoclassical pillars and Renaissance relief sculpture, No. 17 presents a dignified, well-proportioned facade with handsome fluted columns. Two towers rise from the roof and together with twin entrances in the facade, imply a Baroque influence.

No. 17 is so narrow that it's often easy to miss the eight dramatic man-size Atlantes, Atlas-like sculptures that appear to be supporting the roof beneath the cornice. It took Japanese craftsmen five months to carve each sculpture, each requiring three pieces of Italian granite. They are bent and their muscles bulging with apparent effort to hold up the roof.

The editorial offices were on the fifth floor, other floors were leased and the rear seven stories were occupied by the printing press and Linotype machines. A hollow double wall separated the front and rear of the building to muffle the roaring of the presses and clattering of the Linotypes.

No. 17 is also famous for its tenant, the American insurance company AIG (American International Group Inc), which moved into the building in 1928 and used the northern door as its main entrance.

In 1919 American Cornelius Vander Starr, who was later nicknamed the "King of Insurance", founded the American Asiatic Underwriters (AAU) in Shanghai, which became the largest insurance empire in Asia, the forerunner of AIG.

In 1941 No. 17 was occupied by Japanese army and AIG moved to Hong Kong in 1947. However, its former branch company AIA, which also had an office in No. 17 in the 1930s, returned to the building in 1998, becoming one of the first original companies that returned to its old Bund office.

The *North-China Daily News* closed in 1951. Thereafter, state-owned enterprises, including the China Silk Import & Export Co, occupied No. 17.

After living here for six years, Hahn, a famous character in old Shanghai, finally left the city and her married lover forever in 1940. She wrote her famous book, *The Soong Sisters* (1941), went on to have other adventures, marry, have children and write prolifically. She died at the age of 92.

From 1850 to 1951, the century recorded by the *North-China Daily News* was the most breathtaking in the city's history. Legendary figures like Hahn and Starr left their footprints and legends all over Shanghai - and No. 17 on the Bund.

Tips

The ground floor is open. Crane your neck to appreciate the Atlantes sculptures "holding up" the roof.

外滩 18 号 No. 18

菲利普的方案
UNESCO Awarded Building

外滩18号明亮而美丽，但2004年大修前却好像一个低档旅馆。原本气派的大厅被夹层压矮了一半，黯淡压抑。如何修复改造，让意大利设计师菲利普陷入沉思。

建于1923年的外滩18号原是英商麦加利银行大楼，1922年大楼尚在兴建时，英文报纸《远东时报》就报道了这座古典主义希腊复兴风格建筑。

"入口处美好的铜门是英国制造的。方形门厅的四角分别由四根带裂纹的意大利大理石柱点缀，墙面则饰有另一种奶油色大理石，搭配黑色柱基。地面由罗马大理石马赛克铺就，而屋顶镶嵌着瑰丽的石膏天花。"

如今，大楼导览人员对报道中的亮点如数家珍，这些珍贵遗迹历经大改造而得以留存，得益于菲利普的方案。

当年他的团队没有草率开工，而是花费了3个月的时间深入大楼内部研究。他们层层剥茧，发现了大量被后期加建所覆盖的建筑细部。20世纪初的建筑技术让人深深惊叹，如此壮美的建筑居然只用了14个月的时间就建成了。最终，他们推出了一个"半夹层方案。"

菲利普认为如果完全拆除夹层，原本高达7米的银行大堂会相当不实用。他决定仅拆除夹层的中央来恢复大厅的敞亮，同时将夹层两侧改造为二楼商铺，以提高空间利用率。

他们还用威尼斯传统方法清洗了风化的大理石表面，并在内墙使用了一种名为"Mamorino"的专利涂料。乳白沁凉的墙壁摸上去有大理石质感，却没有一丝石块的接缝，同时又散发柔和的光泽，非常神奇。2006年，改造工程荣获联合国教科文组织颁发的遗产保护奖。

"有人主张修复时严格恢复历史原貌，但结果往往是一个'死亡的建筑'，气息陈旧，好像失去生命力的博物馆。意大利人喜欢为老建筑注入新的生命与活力，这就是成功所在。"菲利普如此诠释方案的奥妙。

参观指南

建筑对外开放，门厅里就是带裂纹的意大利大理石柱。

Balancing historic preservation with new commercial functions has always been an issue with the architecture along the Bund.

No. 18 on the Bund, now widely known as Bund 18, received a UNESCO Heritage Award in 2006 for its efforts to achieve that delicate balance, so that it would be a functioning work of architecture, not a museum.

The five-story, granite-faced Neo-Grec structure was completed in 1923 as the office of the Chartered Bank of India, Australia and China. Founded in London in 1853 and opening a Shanghai branch only four years later, it was among the earliest foreign banks in China.

The English-language *Far Eastern Review* reported in 1922 that the 1,755-square-meter, steel-frame structure would be "of heavy, dignified, classic design, Neo-Grec in style with little ornament."

The publication noted a pair of bronze gates made in England, a square vestibule with four Brecchia marble columns, one in each corner, a floor of Roman marble mosaic, and a plaster ceiling.

Today in a standard Bund 18 tour, guides point out "gems" of the building, such as the gates, marble columns, mosaic floor and other features.

Because of its award-winning renovation, these features survived to be appreciated in a new, high-end retail center.

Filippo Gabbiani of Kokaistudios says Bund 18 is remarkable because of the large number of original features that have been preserved. He was commissioned in 2002 to renovate the building for mixed retail use. His team spent the first three months surveying the architecture and then worked two more years, virtually 24 hours a day, to complete the project. It opened in November 2004.

The architect found original finishes on the lobby, the original columns, marble, floorings and decorations hidden under layers and layers. He was amazed that such a magnificent building took only 14 months to build with the technology of its time.

If restoring the historic details is difficult, redeveloping the building on the basis of preservation for viable commercial use is more challenging.

The ground floor was originally a large open gallery with high ceilings for the banking hall, but a suffocating mezzanine had been added in renovation in the 1980s to increase the usable space. In the renovation, the central area was reopened for a bright, breathtaking effect, though the mezzanine on both sides was maintained, connected with the ground floor by way of two symmetrical stone staircases.

Now Bund 18 is a chic lifestyle center filled with restaurants, bars and an art gallery. The mezzanine maintained on both sides housed two lines of luxury retail shops.

"A 7-meter-high lobby for a bank? I was not going to use it in an efficient way. Some people copy all the original decorations inside and they bring it back without tearing anything down. The result is a dead building. It smells old and it's like a dead museum. Italians want to bring life back to the buildings. We succeeded because we bring life back to the building," Gabbiani says of this award-winning "half-mezzanine plan".

Tips

The building is open to the public with restaurants, bars and luxury shops. Note the Brecchia marble columns in the vestibule.

在上海的城市记忆中，外滩19号是和平饭店南楼，但它的历史其实更加悠久，1908年建成时是豪华的汇中饭店。远远望去，大楼好像一座漂浮在外滩的童话王宫，而饭店的英文名"The Palace Hotel"确有"王宫"的意思。

大楼由司各特设计，清水红白砖的立面在灰色的外滩生动夺目。密布繁复雕花的南京路入口还有一扇老式转门。

6层高的饭店有120间客房和1间200人大宴会厅，作为当时的"高层建筑"还安装了2部我国最早的电梯。1909年2月，"万国禁烟会"在此召开，1911年孙中山在这里举行庆祝宴会，1927年，蒋介石和宋美龄在此订婚。

汇中饭店因为有上海最早的屋顶花园而风靡一时。花园由人工草坪和绿色藤蔓装饰，有一个巴洛克式的亭子。在这里边喝咖啡，边欣赏江景和对岸浦东田园风光，是绝佳的享受。

可惜1912年8月的一场大火影响了酒店生意，外白渡桥边的礼查饭店新楼又吸引走不少客人。后来汇中饭店虽一度复兴，但由于更摩登的华懋饭店（今和平饭店）和国际饭店相继建成，终难再现辉煌。

抗日战争期间酒店曾被日军占据，新中国成立后又用作办公，直到1965年改为和平饭店南楼。到了20世纪80年代，"王宫"又差点被拆除。几经波折，昔日的荣耀几乎被忘却。

2010年，外滩19号被改造为斯沃琪和平饭店艺术中心，内部包含名表店和艺术套房。从此其迎来新的命运。

改造工程保留了大木梯和蒋宋订婚厅的原貌。在二楼举办钟表主题展的大厅里，原始立柱和灰色砖墙刻意裸露着，既有历史感，又很后现代。细心观察下还能看出百年前客房并不宽敞的开间布局。

大楼的中间两层空间用于一项有趣的活动——开放了18套工作室供来自全球的艺术家免费使用6个月。该项目由早年热爱艺术创作的斯沃琪集团首席执行官海耶克发起。参加项目的西班牙画家白怀义坦言，其创作的最大灵感源泉竟是南京路喧嚣的夜晚。偌大而安静的画室里，他常常从这外滩"王宫"的券窗向外眺望，那么多闪光灯，还有夜外滩的迷离灯光，后来都凝固在一幅幅斑斓的画作中。

参观指南

酒店对外营业。建议从古老的楼梯拾级而上，到展览厅里可以看到裸露的原始墙壁和柱子，或到100年前上海最著名的屋顶花园喝一杯。

Faced with red-and-white bricks, No. 19 on the Bund looks like a palace from a fairy-tale, perched on the corner of Nanjing Road E. and the Bund. A century ago it was the Palace Hotel.

Widely known as the Peace Hotel South Building, No. 19 has a much longer history than the Peace Hotel building that opened in 1929 as Cathay Hotel.

The six-story Palace Hotel used to be one of the largest, tallest, and most commodious hotels in China, and was an important venue for gatherings of Chinese elite in the International Settlement before 1949.

In 1911, Dr Sun Yat-sen hosted his great banquet for 100 guests at the hotel to celebrate his victory in the presidential election. Kuomintang leader Chiang Kai-shek and Soong Mei-ling held their engagement reception on the top floor in 1927.

The hotel was also the venue of the 1909 International Opium Commission meeting, marking one of the first steps toward banning the trade in opium.

Aiming to compete with the world's best hotels, the Palace Hotel had 120 rooms, a 300-seat dining room, a 200-seat banquet hall and a famous roof garden.

It had city's first elevator, an Otis elevator, installed in 1907.

The grand hotel was also famous for an elaborate roof garden, with a Baroque tower, a pair of cupolas and artificial lawn. Green vines wound around the railings.

It was also the city's first roof garden which offered a bird's-eye view of the city as well as the countryside opposite the Huangpu River. While sipping a glass of whisky on ice, customers could admire the music from the Municipal Orchestra concerts in the Public Garden during summer weekends. On a winter afternoon, they could drink hot coffee under warm sunlight while appreciating the river scene.

Unfortunately, a fire that started on this rooftop on August 15, 1912 damaged the reputation of the Palace Hotel, which lost clientele to the new building of the nearby Astor House Hotel. The hotel had some good times in the late 1910s and 1920s but later suffered from the competition with modern "skyscraper" hotels, such as the Cathay Hotel at No. 20 and the Park Hotel.

The building was occupied by the Japanese during World War II and used by several state-owned organizations after 1949. In 1965 it reopened as the Peace Hotel South Building.

Thereafter, the past glory of the Palace Hotel was almost forgotten. However, a renovation completed in 2010 transformed the old-fashioned No. 19 into the new-concept Swatch Art Peace Hotel, featuring flagship shops of the Swatch Group, a spacious exhibition hall, a luxury boutique hotel and two floors for "Artist in Residence" program.

Each year, dozens of artists from around the world are offered a stay on the Bund for months free of charge. Each receives a room and separate studio.

The 2010 renovation changed the layout but retained the dark-toned wooden staircase. The original interior gray-brick walls and columns are exposed. There's a post-modern feel to the space.

The famous roof garden opens from April to November every year. The Baroque tower has been transformed into a cigar bar while one of the two cupolas is now a dining room.

Today No. 19 still bears the inscription "1906" above the main entrance on Nanjing Road, 1906 being the scheduled completion date.

No. 19 is also one of the only two remaining red-brick waterfront buildings, No. 9 being the other, both standing out along the gray stone structures along the Bund.

Despite ups and downs during the past century, the red brick "Palace" is now a red-hot highlight that enlivens the gray skyline of the Bund.

Tips

The hotel is open to the public. I would recommend climbing the antique staircase to appreciate the exhibition room featuring exposed original walls and columns, or have a drink on the roof garden when it's open.

外滩 20 号 No. 20

摩登的"老贵族"
A Hotel of Memories

"一个老贵族",这是和平饭店给人们的印象。

"他是位老人,但他是个贵族,一个'lord',非常讲究、非常正统。"同济大学建筑系常青教授如此回忆。2007年大修前,他的团队曾对饭店进行了大楼竣工80年来的首次全面测绘。

1929年9月5日沙逊大厦开业,英文《字林西报》刊登的广告十分诱人,称位于大厦内的华懋饭店(和平饭店前身)是"艺术与奢华的完美结合"。

广告并未夸大其辞。饭店著名的九国套房以中、英、美、印等国主题布置,异国情调浓郁。套房内有嵌入式壁橱和大理石浴缸,银质龙头流淌出净化水,舞厅和宴会厅镶嵌着价值连城的法国"拉立克玻璃"。

公和洋行建筑师威尔逊在1925年巴黎博览会结识了拉立克先生。后者设计的这种乳白色玻璃经灯光照射即可闪现蓝光或耀眼的桔红光芒,美丽无比。"拉立克玻璃"后来也成为装饰艺术风格的经典手段。

在标志装饰艺术风格诞生的巴黎博览会上,威尔逊对这种注重装饰、喜欢几何图案且色彩绚丽的新风格兴趣

浓厚。当时他正在酝酿沙逊大厦的方案,原本的设计风格还是新古典主义的,但最终决定紧跟国际潮流,改为装饰艺术化的浪漫新古典风格。大厦建成后不仅成为外滩地标,更将上海全面推向了装饰艺术时代,国际饭店和百老汇大厦等装饰艺术派大楼在数年后纷纷落成。

今日走进和平饭店好像走进一个古旧的澄黄色的梦。八角厅穹顶和紫铜色吊灯都笼罩着半透明的黄玻璃。昏黄的光影中,各色几何图案若隐若现,与酒店里弥散的氤氲香气裹挟在一起,恍如隔世。

1935年美国女作家项美丽到上海后,曾是这间梦幻酒店的常客。在自传《我的中国》中,她深情地回忆了那一段充实快乐的上海生活:"如果不去弗雷兹夫人家做客,我也许就约一位女朋友到华懋饭店吃午餐。我们习惯到大堂先喝一杯,看看能否遇到有趣的男士。"

事实上,建造大楼的犹太富商维克多·沙逊爵士就是一位相当有趣的男士,后来他成为项美丽人生导师般的密友。爵士来自号称"东方罗斯柴尔德"的巴格达沙逊家族,极具商业天赋。今天和平饭店的甜品店就是以他的名字命名的。

爵士喜欢在大厦顶层的公寓举办主题奇异的派对。他爱好广泛,酷爱

沙逊爵士拍摄的项美丽姐妹的肖像,1935
Sir Victor Sassoon's portrait of the Hahn sisters, 1935

摄影,项美丽姐妹一张颇为传神的侧影就出自他手。

在人生关键时刻,沙逊给了项美丽重要意见。1941年他读完《宋氏姐妹》书稿的第一章,直言不讳:"太闷了,快把我闷死了。幸好我已经上床了。要是坐在椅子上读,我一定会坐着就睡着了。"

这则评价与《宋氏姐妹》生动出彩的开篇迥然不同,显然项美丽听取了他的意见进行了修改。这本畅销书也成为这位传奇女子的成名作。

不过,摩登爵士的顶楼公寓还是传统保守的英式风格,毫无装饰艺术元素,一点也不时尚。也许这地道的老贵族的感觉,才是爵士真正深爱的味道。

参观指南

酒店的和平收藏馆有很多旧照片和老家具,值得一看,或到一楼甜品店"Victor's"喝杯咖啡。最好选择坐在窗边,沐浴在清晨阳光之下看南京路的街景。

Every Bund building is a showpiece, but No. 20 is the one that is most closely associated in people's minds with Shanghai itself.

Social change in China has split the history of the former Sassoon House in two parts. As the Cathay Hotel, it was famous throughout Far East since 1929 and set a precedent for luxury and glamour. In 1956 it reopened as the state-owned Peace Hotel, a city landmark.

No. 20 was built on the city's most expensive land, a T-shaped area comprised of part of the Bund road and Nanjing Road E. It was named Sassoon House after businessman Ellice Victor Sassoon (1881-1961), who dominated commercial life in Shanghai for many years.

As the fourth generation of the influential Sassoon family, he served as a captain in the Royal Air Force during World War I, until a plane crash left him with a limp.

The man with a superb nose for business chose Shanghai, where he built apartments, office blocks and hotels. No. 20 was his ultimate showpiece on the Bund built on the site of old Sassoon twin buildings.

In the 1991 book *Shanghai*, British historian Harriet Sergeant described Victor as an "amusing, cynical" bon vivant who held brilliant parties in his penthouse mansion in the Cathay Hotel.

Once he invited his guests to dress as if they had been caught in a "shipwreck" and another time the theme was "circus." So the costumes ranged from pajamas, shower curtain to slippery outfit that resembled a performing seal.

The first to the third floors were offices for Sassoon's company and other enterprises. The Cathay Hotel occupied the ground floor and the fourth to the ninth floors. Sassoon's penthouse occupied the 10th and 11th floors.

Tongji University professor Chang Qing, whose team surveyed No. 20 before its 2007 renovation, discovered the building "was China's first Art Deco building but the original plan was a neo-classical high-rise."

His research showed architect G.L. Wilson from Palmer & Turner was greatly inspired after attending the 1925 Exposition in Paris when Art Deco style was first dated. There he also met famous glassmaker Rene Lalique whose glass he later used to adorn the Cathay Hotel.

Wilson changed his plan to a much more avant-garde building, combining the Commercial Gothic and Art Deco styles that we see today.

Divided by classic three sections, No. 20 is not a pure Art Deco architecture but Wilson uses a lot of Art Deco features, including opalescent Lalique glass. The building also showcases commercial Gothic features, which interestingly echo with the neighboring Club Concordia (demolished in the 1930s to build the Bank of China) in Gothic revival style. Therefore, Professor Chang calls it a "Shanghai Deco" building.

An inviting advertisement announced the opening of the Cathay Hotel in the North China Herald in 1929, noting it was a "wonderful combination of art and luxury" with Lalique glass and lighting and suites in the styles "of all nations."

The advertising was honest. Despite its age and renovations, entering the hotel is like walking into a nostalgic dream.

The centerpiece of the lobby is a spectacular domed rotunda adjacent to an arcade. The dazzling stained glass in the rotunda, elegant metal lamps and abundant Art Deco carvings keep visitors lingering in the warm, yellow tones of the lobby. A soothing wood-scented aroma coming from the original 1929 Art Deco air-conditioning rents fills the air.

As a close friend of Sassoon, legendary American writer Emily Hahn, author of *The Soong Sisters* frequented the hotel.

"I might meet a girl for lunch at the Cathay, with drinks first in the lounge; that meant we would pick up men and make a party of it," she writes in her autobiography *China to Me*.

The hotel had 200 rooms and nine famous "themed" suites, each decorated in a distinctive national style, including Chinese, Indian and English. Each suite had built-in wardrobes. The bathrooms contained marble baths with silver taps and purified water. The hotel's dining rooms were decorated with colorful, blazing Lalique chandeliers.

The well-preserved Sassoon residence itself was decorated in Jacobean style, with dark, carved paneling and richly molded ceilings, instead of the up-to-date Deco.

The greyhound image, which appears at the top of the Sassoon family coat of arms, appears here and there in the former Cathay Hotel. The greyhound symbolizes courage and loyalty.

The hotel impressed Prof. Chang as being an old aristocrat. "He was aging but he was still a lord. Everywhere, every detail is so refined and stylish," Chang says.

Sassoon's parties came to an end with the outbreak of the World War II, and he left Shanghai in the spring of 1941. The building was used by the new municipal government after 1949 until it reopened as the Peace Hotel in 1956.

The state-owned hotel was one of the city's few accommodations for foreign guests before international hotel brands entered China. During the planned economy period before 1978, the hotel did not receive individual guests but only accommodated those who booked through a state-owned organization, such as an export company or a travel agency.

The hotel had no sales department until the business began to decline as those companies retreated from the Bund. After 1978, the hotel hosted more foreign tourists and businessmen. The rate was only at a four-star level due to its aging facilities.

After a three-year renovation, the building reopened in 2010 as the Fairmont Peace Hotel. Today, it is again a hotel of art and luxury on the Bund.

Tips

The Peace Gallery exhibits some historic archives of the hotel. I'd also suggest have a drink at window-side table of the "Victors' café" with a view of the Nanjing Road and usually with nice morning sunlight.

外滩 23 号 No. 23

外滩的斗拱
"Dougong" on the Bund

外滩一行洋楼中,只有23号中国银行带有浓郁的中国风。这座中西合璧的银行是由公和洋行威尔逊和中国建筑师陆谦受共同完成,也是外滩唯一一座华人参与设计的临江大楼。

这座西式摩天楼的中式元素可谓琳琅满目,如四方攒尖屋顶、铜绿色琉璃瓦和镂空花格窗。特别是檐部竟以石质斗拱来装饰,让人惊叹中西文化可以如此水乳交融。

中国银行的前身是清政府的国家银行——大清银行。1912年南京临时政府组建中国银行,办公地先设在汉口路50号大清银行旧址,又在北京成立总行,1928年从北京迁回上海后选择在外滩原德国总会营业。

但总会建筑究竟不适合业务蒸蒸日上的中国银行,总经理张嘉璈认为"银行实力足以与驻在上海的欧美银行相抗衡,必须有一新式建筑,方足象征中国银行之现代化,表示基础巩固,信孚中外"。为筹建新楼,1934年银行曾搬回汉口路旧址过渡。

1936年10月10日,银行举行新楼奠基礼,隆重的会场由翠柏彩花和五色布幔装点。据《密勒士评论报》报道,总经理宋汉章的致辞深情回顾了历史:"1912年2月中国银行在汉口路50号成立,两年前我们又为建新楼而搬回汉口路办公,在不到四分之一个世纪的光阴里恰好完成一个完美的循环。而新楼的崛起则意味着自此我行将告别过去,开创新的篇章。"

银行原打算建造一座34层的巨厦来展示自身实力,但"雄心大楼"的高度后来大大缩水,个中缘由成了外滩的著名谜团。犹太富商沙逊因担心高度超越沙逊大厦而向工部局施压是常见版本,中行档案1934年的会议记录也显示了管理层对工部局反对的担忧,但也有专家认为不景气的经济才是高度缩水的主要原因。

无论如何,最终的结果对于外滩倒是一件幸事。气势凌人的巨厦本来将破坏外滩美丽的天际线,而修改后的方案与万国建筑群十分和谐,中国元素也更适合一家中资银行。不禁想到邬达克的作品档案中保存的一份外滩9号招商局巨厦的设计草图。倘若当时那座充满未来感的40层巨厦建成,今日的外滩又是另一番景象了。

参观指南

中国银行一楼在营业时间对外开放,楼内有个免费开放的行史博物馆,需要预约(预约电话:021-63292618)。

In the 1930s, the Bund that we see today was completed and the Bank of China was one of the last few strokes on the beautiful skyline.

No. 23 on the Bund, the Bank of China building, stands out as a rare example of Chinese Art Deco on the mile-long Bund dominated by a range of Western-style architecture.

Completed in 1937, it is the only heritage building on the waterfront that was invested by Chinese, co-designed by Chinese and adorned with abundant Chinese elements.

The white building was originally designed to stand 34 stories, around 100 meters, which would be China's tallest building at the time. But the plan had shrank drastically to a 17-story, 70-meter-high structure, which remained one of the most famous myths lingering on the Bund. It is widely said that magnate Victor Sassoon, owner of the 77-meter-high No. 20 next door, insisted that the new building should not be taller than his Sassoon House.

The Bank of China was founded in 1912 by the Kuomintang in Shanghai as the continuation of the former Da Ching Bank, the Qing Dynasty's (1644–1911) royal bank, which closed soon after the 1911 Revolution.

Initially, the Bank of China used the Da Ching Bank office at 50 Hankou Road. The head office now was formally established in August 1912 in Beijing where it remained until 1928 when it was moved back to Shanghai to the Club Concordia at No. 23 on the Bund.

As the bank expanded swiftly along with Shanghai's booming economy, the club building could no longer accommodate all the business.

"The power of the Bank of China could compete with American and European banks. We had to build a new-style building to symbolize the modernization of the bank, to showcase our solid foundation and superb credit to China and the world," the bank's president Zhang Jia'ao (Chang Kiangau) said at the time.

The original and ambitious blueprint called for a powerful, soaring structure, similar to the American Radiator Co Building in New York.

It would have towered over Sassoon House and the tycoon is believed to have influenced the Shanghai Municipal Council to deny a building permit for such a tall structure.

The record of a bank meeting in 1934 indicated that top managers were concerned the council might oppose the height.

But other documents showed that Shanghai's declining economy also led to the later reduction of floors.

And if the "dominant" skyscraper had been built, its height and impeding style would have ruined the harmony of buildings along the Bund. The revised plan is a better one, which has added more Chinese characters for this Chinese bank and looked compatible with its neighbors on the Bund.

The revised East-meets-West blueprint was also an East-West cooperation co-drafted by George Wilson of Palmer & Turner and British-trained Chinese architect H. S. Luke (Lu Qianshou).

Built on a 5,110-square-meter plot, the 17-floor structure is a Western Art Deco skyscraper graced by Chinese elements inside and out. It has a simple, square Art Deco body and a slightly overhanging Chinese roof of glazed tiles supported by stone dou gong or typical Chinese supporting brackets.

The main body of Chinese granite is dotted with lattice windows, similar to lattice windows in traditional Chinese wooden structures. Chinese cloud patterns are repeated here and there on beams and columns.

The building was a well-equipped modern office, fully air-conditioned and serviced by 13 elevators. It contained a grand banking hall lined by 16 black marble fluted columns.

Despite the reduction in scale, No. 23 stands out proudly among the Neoclassic and Gothic revival Bund architecture as a powerful and stylish statement of China.

As the oldest surviving bank in China, the 103-year-old Bank of China has continuously used the building, which it occupies today. The bank specialized in foreign exchange and international trade after 1949. Today it is one of the four large state-owned banks. Back on October 10, 1936, the corner stone of No. 23 was laid in ceremony attended by many Chinese dignitaries and foreign guests. The ceremony took place in a temporary structure, decorated with flags and closely guarded by police.

T. V. Soong, chairman of the bank's board, delivered an inspiring speech.

20世纪30年代高度超越沙逊大厦的原设计草图
The Original Sketch of the Bank of China Building in the 1930s

"In February 1912 the Bank of China was organized with our Head Office at 50 Hankow Road, Shanghai... Two years ago, when the old building here was about to be demolished, the Head Office and the Shanghai Branch were retransferred to 50 Hankow Road, thus completing a perfect cycle in less than a quarter of a century. The erection of a new building seems to indicate that the Bank is about to close one chapter and open another," he said.

Tips

The bank is open to the public with an informative history museum inside No. 23, which is free but requires reservation at 021-6329-2618 (Chinese only).

外滩 24 号 No. 24

邬达克的外滩工作室
Hudec's Atelier on the Bund

1926年夏天，国际饭店设计师邬达克亲笔书写了一份傲人的简历："1925年1月我创立了自己的公司，此后便设计建造了宝隆医院、西门外妇孺医院宿舍楼、宏恩医院……"

邬达克的创业地就在外滩24号横滨正金银行新楼，该建筑曾被《远东时报》形容为"一颗新嵌到外滩皇冠上的珍宝"。

这又是一件公和洋行的作品，建筑师弗兰克·科勒德的灵感新鲜而有活力，巧妙结合了日本元素和新希腊复兴风格，让"东西方风格和谐融为一体，而不像其他中西结合建筑给人以凌乱的感觉"。

外立面浅色的日本花岗岩与黑色铁门形成鲜明对比。窗下的锁石形似低垂双目的古佛，当年的报道幽默点评："这也许表现了现代银行家淡定的一面。"一楼室内墙上，装饰有青铜雕刻的古代日本武士，身披盔甲，肩背箭羽。

建造24号的这家日资银行历史悠久，1880年在日本横滨成立，1893年即开设上海分行。1924年7月大楼开业仅几个月，年轻的邬达克便登报宣布在此开设事务所。在24号工作的7年里，他的灵感源源不绝，完成了四行储蓄会大楼和慕尔堂等代表作，并着手设计大光明电影院和国际饭店。

24号也是一座注定与银行结缘的大楼。1945年该楼改作国民党中央银行，1949年成为中国人民银行华东区分行，2000年工商银行上海分行迁入办公。

今天的银行大堂依然气派，椭圆形穹顶由金线勾勒，8根灰白色大理石方柱分列两侧。这个美丽的玻璃穹顶引入充足光线，照亮了每个角落。

《远东时报》曾经评论："让人印象深刻的不仅是面积，而是这里不像其他银行有难看的铁栅和华丽铜饰。在横滨正金银行，业务多在宽大的抛光硬木柜台上进行，只有现金柜台才使用铁栅。"

如今24号大厅里也有宽大的开放式柜台，点着一行柠檬色光芒的古铜台灯。光影交错间，仿佛还是间90多年前新落成的银行。而楼上，一位新创业的建筑师在异乡勤奋地绘图。在这外滩珍宝里尽情创作的他，也许已萌发超越外滩建筑的梦想与雄心。

参观指南

工商银行一楼在营业时间对外开放。请注意这座西式建筑上的东方元素，如青铜日本武士雕像等。

No. 24 was called a new jewel on the waterfront when it was completed in 1924.

"Shanghai's crown of jewels is her Bund and the latest, if not the brightest ornament therein, is the Yokohama Specie Bank building, into which more than two score of manufacturers have poured their cornucopias of the best in modern building materials to make this structure unique," the English-language *Far Eastern Review* reported.

Designed by Frank Collard from Palmer & Turner, this structure is where legendary architect L. E. Hudec kicked off his own business and later earned international recognition for designing the Park Hotel and many other buildings.

No. 24 was built to house the Japanese Yokohama Specie Bank, the earliest "Shanghailander" among around 10 Japanese banks that had set foot in the city, which opened its doors in Shanghai at Nanjing Rd E. in 1893 and later moved to No. 21 on the Bund. As the growing business required greater space, it purchased the plot at No. 24 for a new Shanghai branch.

Architect Collard displayed freshness and vigor in this building, adapting Japanese features to the general Neo-Grecian exterior design in a seamless way. The result is East and West in harmonious unity, without those messy lines that commonly crop up in buildings where an attempt is made to graft West upon East or vice versa.

On the exterior, Japanese granite forms a fine contrast with the black iron gates. Bronze sculptures depicting the helmet and arms of ancient Japanese warriors grace the side entrance hall. They are fitting guardians for a financial institution.

The plain granite facing is continued throughout the interior of the spacious portico, which serves as the base of the classic Ionic columns that run up to the main entablature.

The effect of Oriental touches is heightened by the keystones of the window arches. The sculptured granite Buddha heads with downcast eyes "mayhap express only the calm resignation of the modern banker."

Inside, the building is not only flooded with light from ample windows, but also is illuminated from above by a beautiful glass dome that sheds ample daylight into every corner of the great room.

Only half a year after No. 24's grand opening in July of 1924, Hudec opened his own studio here.

During the seven years at No. 24, Hudec designed many of his signature buildings including the Moore Memorial Church, the Grand Theater, and the Park Hotel.

No. 24 was also destined to house financial institutions, and most of its owners were banks. After World War II ended in 1945 the bank was taken over by the Kuomintang government's Central Bank.

After 1949 The People's Bank of China relocated its eastern China branch here before the city's textile industry bureau moved in. China's largest bank, the Industrial and Commercial Bank of China, launched its Shanghai branch at No. 24 in the 1990s.

Today historical details are preserved or revived, including the Buddha-head keystones, the bronze sculptures of Japanese warriors, the wide staircases with handsome balustrade in black marble, and eight huge gray marble columns below a golden glass dome in the banking hall.

It mirrors the impression of a newspaper reporter who entered the building almost 90 years ago. "It is impressive not only in size but in the absence of much of the grill work and ornamental brass that disfigures many modern financial institutions. In the Yokohama Specie Bank most of the business is done over the broad polished hardwood counters and only at the cashier's end is grill work used at all."

Today the banking hall also features a line of open-style broad marble counters on the right side. Lighted by a string of antique lamps that cast a warm lemon light, for some moments the banking hall looks again a newly completed, turn-of-the-century building that had seamlessly put together Eastern and Western features. Upstairs a homesick architect was working hard on his drawings in the remote foreign land. The architecture sprang from his imagination later had left an indelible mark on the city's landscape.

Tips

The banking hall is open to the public. Please note the Oriental features in the western building, such as the bronze sculptures of Japanese warriors.

外滩 26 号 No. 26

保险大楼与外滩风景
Monument to Early Preservationist

罗马不是一天建成的,外滩也不是。外滩26号的故事不多,却与一个改变外滩命运的美国人相连。

大楼建于1918年,由扬子保险投资。这家保险公司来头不小,1862年由美商旗昌洋行大班金能亨发起创办。当时洋行垄断了长江航运,开设保险业务可减少事故损失。

在早期租界,金能亨是有份量的人物。他兼任美国驻沪副领事,又是工部局1854年创立时的首任董事,1868年升为总董。

工部局会议记录中珍藏着一封金能亨于1869年底写自日本横滨的信,当时刚刚卸任总董的他在信中强烈反对继任者将滨江大道用于建造码头仓库的计划:"外滩是上海唯一的风景点……这是城中居民黄昏漫步时唯一可以呼吸新鲜空气的地方,也是租界中唯一有开阔景色的地方……航运业并非商业的主要元素,而只是其低等的附属行业之一,类似驮马和载重马车。而交易所、银行、清算所才是商业的中枢,它们所在地总是商业最繁荣的地方。"

这位富有远见的美国人认为航运业的噪声灰尘将吓跑金融机构,千万不

能把上海变成"另一个乌烟瘴气的利物浦"。卸任的总董余威仍在,这封长信也改变了滨江大道即将沦为码头的命运。

1891年旗昌洋行倒闭,扬子保险独立营业,鼎盛时期开设了30多家分号。26号大楼高达7层,英文报纸《远东时报》在其竣工前就预告"又一个美丽的建筑即将在外滩崛起"。

大楼由公和洋行设计,呈新古典主义风格,看似中规中矩,细节却富有变化。底层为平梁式入口,二层变化为半圆形券窗,壁柱通贯三到五层,六层中部是双柱柱廊,屋顶花园视野绝佳。

扬子保险曾被日军接管,二战后复业生意不错。大楼吸引了多家保险公司前来租赁,成为一座名副其实的"保险大楼"。1950年后公司结束在华业务,中国粮油进出口公司等单位进驻26号办公,近年又改造为农行营业厅。1931年到1957年,丹麦王国驻沪总领事馆曾设于三楼。

一个多世纪前,喜爱外滩的人们与想建码头仓库的洋行不懈地斗争,几乎每隔几年就要爆发一次,直到1900年前后保留滨江大道的观念才成为普遍共识。

法国著名史学家白吉尔在《上海

史——走向现代之路》一书中描绘了1900年外滩商业区的模样——"银行、房地产公司、保险公司和奢侈品店林立"。

"这个区域的重点是金融与服务。当时人口密度是每平方公里60 000人左右,虽然人口停止了增长,但是地价持续上升。在滨江大道边,石材或混凝土建造的大厦展示了大公司的财富和实力。"

也许,这个1900年的外滩正是31年前金能亨在横滨不眠夜里构想的样子。他或许没有想到,亲手创建的扬子保险会在外滩商业区拥有一座展示实力的大厦。

了解这些故事后,漫步外滩,或在一个有风的露台眺望滨江风景,不由暗暗地感谢金能亨们的努力。

而今日外滩,也许比他们想象的还要美丽。

参观指南

农业银行一楼在营业时间对外开放。

Rome wasn't built in a day, nor was the Bund. The willowy building No. 26 is linked to a man who had saved the beauty of the Bund.

Constructed between 1918 and 1920 for the Yangtze Insurance Association, one of the earliest and largest insurance companies in China, No. 26's slim body features an austere style with refined details.

Yangtze Insurance was founded in 1862 by American merchant Edward Cunningham to insure the cargoes and vessels of the American firm Russell Co's Shanghai Steam Navigation Co, which dominated shipping on the Yangtze River.

A Russell partner and US vice consul in Shanghai, Cunningham was one of the seven board members when the Shanghai Municipal Council of the International Settlement (SMC) was created in 1854. Elected chairman of the board for 1868–1869, the man is remembered today for his vision for the Bund.

Shanghai Archive Bureau preserves a letter in the meeting minutes of SMC archives, in which Cunningham strongly opposed the plan to use the river frontage of the Bund for wharves.

"The sole beauty of Shanghai is the Bund... It is the only place where the residents can get fresh air from the river in an evening promenade, the only place of the settlement where there is a free outlook," Cunningham wrote in the letter. It was dated on December 30, 1869, from Yokohama, Japan, shortly after he retired from the position of SMC chairman.

"Shipping is not the main element in commerce. It is one of the courses adjunct like pack-horses and drays. The Exchange, the Banks, the Counting houses hold the basics of commerce and where they are located there always is the best quarter, the greatest throng of commercial life," wrote this far-sighted man.

In his opinion, the presence of shipping will bring only noise and dusts, which will scare off those financial institutions. Though retired, Cunningham was still influential and this four-page letter had affected opinion of the SMC board.

Russell & Co went bankrupt in 1891 but Yangtze Insurance was reorganized as an independent company. At its peak, the company had more than 30 branches in Chinese port cities, specializing in maritime and fire insurance.

Covering a land area of 639 square meters, the seven-story, steel-and-concrete No. 26 was reported as "another handsome structure that is being erected on the Bund" by the *Far Eastern Review*.

The edifice was topped by a roof garden that was 115 feet (34.5 meters) above the street level.

Yangtze Insurance Association was taken over by the Japanese after 1941 and resumed business after 1945. In the 1940s, No. 26 also attracted nearly 10 insurance firms that rented space, hence, it was known as the "plaza of insurance."

The nearly century-old insurance company ceased business in China in the 1950s after which state-owned companies including China Oil & Foodstuffs Corp moved in. From 1931 to 1957, the Royal Danish Consulate

General in Shanghai was located on the third floor.

Today, No. 26 is used by the Agricultural Bank of China, which uses the ground floor as a banking hall and upper floors as private banking club and Bund Art Center for the China National Academy of Painting.

A recent renovation has changed the interiors of the ground floor but major rooms in the above floors have maintained the original style, "adorned with expensive imported teak wood" and "contrasting to the simple granite on the external walls," according to *Far Eastern Review*.

Back to more than a century ago, Bund aficionados like Cunningham continued to fight with powerful trade firms that wanted to use the Bund frontage as wharves and docks for ships and warehouses for cargo.

These arguments had arisen every several years until it was widely recognized in the 1900s that the Bund should maintain the promenade and sightseeing functions. No one proposed ship wharves any more.

In the book *Shanghai-China's Gateway to Modernity*, French historian Marie-Claire Bergere describes the Bund area as a business quarter by 1900 where "banks, real estate companies, insurance companies, and luxury shops were located."

"In this quarter devoted to finance and services, where by 1900 density was already over 60,000 inhabitants per square kilometer, the population had ceased to grow, but land prices continued to soar. Along its avenues, stone or concrete buildings testified to the wealth and power of the great companies."

20 世纪 30 年代的扬子保险
Yangtze Insurance Building in the 1930s

The 1900s Bund was exactly as Cunningham had imagined 30 years before while drafting that important, four-page letter during a sleepless night in Yokohama. No. 26 built for the company he had founded would join the ranks of Bund buildings showcasing the power of commerce.

Walking through the riverside promenade or breathing the fresh air on a windy balcony with a Bund view, we cannot help but thank those who fought for the character of the city. The beauty of the Bund today might even exceed their imaginations.

Tips

The ground floor is open to the public. Please note the handsome marble entrance, contrasting with the severe granite exterior facade.

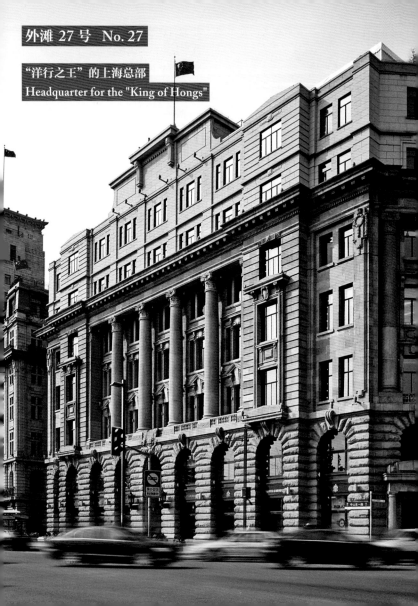

外滩 27 号 No. 27

"洋行之王"的上海总部
Headquarter for the "King of Hongs"

百年间，两家历史悠久的豪门先后来到外滩27号，在这条滨水长廊取得醒目的一席之地。

大楼建于1922年，是"洋行之王"怡和洋行的新楼。洋行由两位苏格兰人创办于1832年，他们以贩运鸦片、丝绸和茶叶起家，最终构筑了一个商业帝国。怡和洋行曾经极力推动鸦片战争的爆发，深刻地影响过近代中国的内政与外交。

1851年建造的怡和老楼拥有壮观的列柱门廊，曾被《弗莱彻建筑史》收录在册。1920年因业务发展极好，洋行决定在原址建造新楼。现代文艺复兴风格的新楼高5层，宽阔的立面上，粗朴的石材基座与石柱营造了坚实稳定的视觉效果。8扇巨大的拱门一字排开，直达二层后自然形成二楼的落地拱窗。

今日走进27号，耳边立刻响起维瓦尔第《四季》中轻快的弦乐，与古旧精美的木门十分搭配。米白色大理石楼梯通往二层的酒窖。

大楼现在是"罗斯福公馆"，设有高档餐厅和奢侈品商店。罗斯福基金会原本计划在此引进美国第五大道百货商店，不料后者中途退出，遂亲自经营大楼。

在德尔·罗斯福主席的眼中，27号是外滩最美的大楼，与其家族地位相称。这位老罗斯福总统的曾孙最喜欢"透过巨大拱窗洒进的阳光"。

当罗斯福的后人们在拱窗边享受外滩阳光时，27号原来的主人就在附近。1979年怡和集团重返阔别25年的中国大陆，上海办事处就设在不远处的外滩中心。

1920年1月怡和洋行宣布建造新楼，高调地将图纸刊登在《远东时报》上。但报道的字里行间充满怀旧之情，因为洋行的老楼曾见证了中国第一条铁路的酝酿，经历了仅有6人的办事处发展为一间大洋行的过程。

"这家在上海发展中扮演如此重要角色的公司将会有一座与之相称的丰碑式建筑，目前正在建设中。但是很多人会遗憾老楼的拆除。那座建筑对于上海的过去真是太有意义了。"

光阴如梭，这段感性的文字落笔已近一个世纪。百年前的外国报人肯定难以想象，怡和新楼的巨大拱窗又见证了多少百年豪门与上海这座城市的故事。

参观指南

建议到二楼酒窖巨大的拱窗前喝一杯，或到视野绝佳的顶楼观景。

No. 27 has housed two historic family enterprises of immense power and wealth, from yesterday until the present.

Today, it is The House of Roosevelt, a commercial property managed by Roosevelt China Investments Corp. Portraits of both Roosevelt presidents (Theodore and Franklin D.) and Teddy bears (named after Theodore) are everywhere in the building, a reminder of the legendary American family.

Back in 1922, five-story No. 27 was erected as the new office for Jardine, Matheson & Co or EWO (Yi He) in Chinese, at one time the largest trading company in China and the East.

Co-founded by Scotsmen William Jardine and James Matheson in Canton (today's Guangzhou) in 1832, the company profited from tea, silk and cotton but its profits came overwhelmingly from the smuggling opium from India into China.

The company had lobbied for years to force China to open up the opium trade, and finally succeeded in persuading Foreign Secretary Lord Palmerston to wage two Opium Wars in the name of free trade.

With a network of wharves and warehouses in major port cities, the company's business expanded to insurance and shipping and in the early 20th century Jardine Matheson became an empire. In 1876, the company constructed China's first railway line.

After purchasing Lot No. 1 on the Bund, Jardine Matheson constructed an office building in 1851, a landmark in its day. The old building with an inviting porch was demolished in 1920 to make way for the new building.

Today, No. 27 gives the impression of a powerful structure, built of reinforced concrete and faced with granite.

Past a black cast iron gate, visitors enter the lobby where a chamber music ensemble may be playing Vivaldi's *Four Seasons*. This leads through an elegant hall to a very large and stately main staircase. The music echoes with the architecture and makes it especially pleasant to explore the building.

Designed in a free rendering of the modern Renaissance, the facades of No. 24 consist of dignified colonnades running through three stories supporting a heavy entablature, the whole resting on a heavily rusticated base.

No. 27 is now a complex of diversified businesses, including luxury retail shops, Roosevelt's private members-only club and high-end dining.

Roosevelt's original plan aimed to establish Saks Fifth Avenue store on the Bund in building No. 27. Saks withdrew from the cooperation.

"To our eyes, it was the most beautiful building both inside and outside. We love the kind of construction and architecture, the enormous windows, the sun and everything," says Tweed Roosevelt, great-grandson of former President Theodore Roosevelt and chairman of Roosevelt China Investments Corp, who thought it was appropriate to represent the Roosevelt brand here with the premier building from the old days.

While Roosevelt was enjoying the Bund's sunlight through the arched windows, the Jardine Matheson Group returned to the Chinese mainland in 1979 after an absence of 25 years and its Shanghai office is in the Bund Center, just 10 minutes' walk from the former headquarter.

In 1920, when Jardine Matheson announced the plan to build a new office, the *Far Eastern Review* published a report filled with memories of the old 1851 building, where the plans for China's first railway were formulated and the Shanghai branch had grown from half-a-dozen employees to a huge organization.

"The firm which has played so great a part in the progress of Shanghai and will continue to hold its place in the future of China will fittingly be housed in the monumental structure which is now under way, although many China residents will view with regret the demolition of the stately old building which has meant so much to Shanghai of the past," the newspaper wrote.

Times flies and it is hard to believe that nearly a century had slipped away since this sentimental news item was written. Twists and turns, retreats and returns, as well as relics of the two foreign family enterprises, all crystallized in the stone building with huge arched windows.

Tips

I would suggest you have a drink behind the arched windows at the cellar or climb rooftop to enjoy a breathtaking view.

外滩 28 号 No. 28

平凡的惊艳
Symbols and Refined Details

与一街之隔的27号怡和大楼相比,同年落成的28号虽然平凡多了,但也有惊艳的细节。

大楼由英商格林邮船公司所建,文艺复兴风格,立面线条干净利落,室内设计的突破在于间接采光。电灯藏于青铜碗状的灯罩内,经天花板反射后的光线非常柔和。而打动建筑师的正是这迷人的光线。

"六楼的窗户镶有彩色的小块铅条玻璃。墙壁很厚,窗户小,透进来的光线有一种特别的历史感。"负责改造工作的现代集团建筑师邹勋回忆道。

改造过程中,在楼梯边的墙壁上发现了一些釉面砖,布满细腻的乳白色龟裂纹,背面印着"产自英格兰"的字样,与釉面砖搭配使用的还有很别致的墨绿色腰线。

在关于上海的城市记忆中,28号就是"广播大楼"。因为1951年后上海人民广播电台和上海市文化广播影视管理局曾先后在此办公。在娱乐生活匮乏的年代,从大楼里传出的电波对于百万上海家庭是重要的精神补给。改造竣工后,央行上海清算所成为28号的新主人。

面向外滩的东门已多年未开,改造后设计了一个英式转门,可以将来宾引入举办鸡尾酒会的大厅。如今一至五层为办公室,六层设有贵宾室,最华丽的601房间保留了原始的柚木墙裙。同层的内天井则改为一个怡人的庭院。

1922年的28号也是设备一流的办公楼。据《远东时报》报道,邮船公司在一层营业,其余楼层用于租赁,每间办公室至少有一个盥洗室。

28号与许多外滩建筑一样是水刷石饰面,这种传统工艺用石屑和水泥等建材塑造出天然质感的墙面,既美观坚固,又耐用经济,20世纪初一度风行沪上。

20世纪50年代,大楼的水刷石曾被部分更换。改造团队发现初建时的水刷石色泽偏暖,而50年代因工艺变化而增加了冷色调。最终采用粘结方法调和冷暖色调差异,让两种痕迹并存。远看无差别,近看却能觉察到微妙变化。

如今修复工作早已结束,但这座貌似平凡却暗藏美丽玄机的大楼之谜仍在年轻建筑师的心中盘旋,特别是601室墙裙上雕刻的四枚符号——六角星、十字架、葡萄和新月。这些符号寓意何在,与邮船大楼的历史又有何关联?建筑师遍查档案也没有找到答案。萦绕外滩的谜团,又多了一个。

参观指南

建筑不对外开放,但可以观察外立面不同时代水刷石颜色的细微差别。

No. 28 is less imposing compared to the neighboring No. 27, which housed the powerful Jardine Matheson. But this more modest building is filled with refined details and myths.

No. 28 was built in 1922 by British shipping company Glen Line Ltd, which was formed in 1910 and amalgamated with the Shire Line in 1920.

The 27-meter-high edifice was designed by Palmer & Turner in free Renaissance style with clear-cut lines. A breakthrough in No. 28 was the use of indirect lighting. The electric lamps had been concealed in large bronze bowls suspended from the ceiling and the pleasingly gentle light had been obtained by reflection.

It was also the special lighting that impressed architect Zou Xun, who was in charge of the building's recent renovation.

"The sixth floor has huge windows adorned by stained glass in small pieces. The walls were thick so the light cast through the small pieces of stained glass gave a touching feel of history," recalls Zou of his first exploration in 2011, who works for Shanghai Xian Dai Architectural Design Group.

In 1951, the Shanghai People's Radio Station moved into No. 28 and its broadcasts were an important part of city life for many years during the era radio was a major entertainment medium. Many programs were produced and broadcast from No. 28 to millions of Shanghai families.

When the radio station moved to Hongqiao Road in 1996, the building was used as the offices for the Shanghai Municipal Administration of Culture, Radio, Film & TV. Its latest renovation houses a new organization, the Shanghai Clearing House of the Central Bank of China.

The ground floor now comprises two halls after the renovation- the southern hall for staff entrance and the grand eastern hall for cocktails and balls. The former gate on the Bund, which had been closed for decades, will be reopened for special events. Zou's team has designed an elegant British entrance hall with a copper revolving door.

Offices now occupy the first to fifth floors. Most rooms on the sixth floor have been preserved as VIP rooms. The same floor also houses a kitchen, a restaurant and a gym for employees. An inner yard, which had been used for haircuts and massage in the days of the state-owned radio station, has become a breathtaking garden.

Back to 1922, No. 28 was also a comfortable modern office. The *Far Eastern Review* reported offices of the Glen Line Eastern Agencies occupied the ground floor while the rest were leased out. The building is fireproof and heated throughout by low-pressure hot water. Modern plumbing and sanitary facilities were installed, so that each office had at least one lavatory.

Zou also fell in love with the ceramic tiles that graced the walls along the staircase.

"The creamy white tiles had subtle crack patterns that perfectly match the emerald belt line tiles in geological patterns, it's very Art Deco," says architect Zou.

The facade of No. 28 is composed of a hybrid of the original 1920s "Shanghai Plaster," which creates the effect of natural stone texture by using small pieces of granite. It is durable, waterproof, economic and beautiful in a solemn way, which has been widely used from the 1900s to the 1930s and regarded as wisdom of Shanghai.

On the facade of No. 28, the 1920s plaster gives a warmer tone than the 1950s plaster, which used more cement. Zou's team has applied a coherent method to mediate their differences, which look the same from afar, but show the traces of different periods from a close look.

The most puzzling discovery occurred in No. 28's most magnificent Room 601. The walls of this former master office are paneled in teak wood that is ornately carved, creating an effect of richness. The teak walls are inlaid with four square frames that contain a crescent, a hexagram, grapes and tiny crosses, respectively.

The architect has not deciphered the symbols yet. Like other Bund buildings, mysteries from the last century are still lingering in places like Room 601, in the seemingly modest, but meticulously detailed No. 28.

Tips

The building is not open to the public but you may admire the nuances in the various types of "Shanghai Plaster" on the facade.

外滩 29 号 No. 29

追逐"华尔街"的法国银行
The Only French Elegance on the Bund

外滩23幢临江建筑中,唯独29号有法资背景。1912—1913年间,法国东方汇理银行翻建了这座精致的小楼。

这家法国银行的上海分行成立于1899年,原在法租界的洋泾浜营业,选择到公共租界的外滩建造新楼,应该是缘于外滩对金融机构的吸引力和品牌效应。在"远东华尔街"上用一座建筑来代表自己,是高明的商业策略。

外滩在19世纪晚期成为金融中心。熊月之在《上海》一书中提道,19世纪90年代起,英国银行独霸的局面被打破。到1927年,35家外资银行设立了上海分行,多在外滩一带。其中,俄资华俄道胜银行、德资德华银行、日资横滨正金银行和东方汇理银行各有一座江景绝佳的办公楼。

法国银行的"外滩旗舰店"由通和洋行设计并施工,呈古典主义风格,立面构图精美。面向外滩一侧的立面,其二、三层中部有两根贯通两层的爱奥尼柱。整座建筑设计的亮点是位于二层正中的窗,三个窗洞采用经典的帕拉第奥母题——中间窗楣上方呈券形,高而宽;两侧为矩形,低而狭。那是16世纪意大利建筑师帕拉第奥的经典构图,灵感来源于罗马帝国时期的建筑型制。

门厅的两侧曾是经理办公室。营业大厅里,柚木基座的六根柱子支撑起石膏天花板和拱形天窗。

东方汇理银行是一家法国政府特许设立的海外殖民地银行,在云南的贸易金融中扮演过重要角色——曾为滇越铁路建设募资。

1955年银行结束在华营业,外滩29号由市公安局使用。20世纪90年代后,上海政府将国有企事业单位陆续迁出外滩,重新吸引金融机构入驻,恢复这条昔日"远东华尔街"的繁盛。光大银行上海分行作为首家国内金融机构成功置换进驻29号。

而29号昔日的主人其实也回来了。银行保留了老上海时代的中文名"东方汇理",1997年从延安路迁往国际金融中心,就位于上海的"新华尔街"——陆家嘴。这一追随发展机遇的举动,与百年前在外滩筹建新厦的决定何其相似!

参观指南

光大银行一楼在营业时间对外开放。请体察这座小楼外立面极为和谐的比例,还有建筑师对古典柱式的巧妙运用。

No. 29 is the only one built with French capital among the 23 heritage buildings on the Bund waterfront. The low-rise, three-story bank in Classic style was built for the French bank, Banque de L'Indo-Chine.

Established in Paris in 1875, the bank had financed France's commercial activities in its Asian colonies. In 1911, the bank purchased the lot to build the structure moved from the former French concession to join China's "Wall Street," the congregation of international banks and institutions in the former international settlement run by the British and Americans.

The Bund was the city's No. 1 show window for financial services which became a financial center in the late 19th century.

Historian Xiong Yuezhi's book *Shanghai* describes how the monopoly of British banks in China was broken since the 1890s, and by 1927, a galaxy of 35 foreign banks set up Shanghai branches, mostly around the Bund, including Russia's Russo-Chinese Bank (No. 15), Japan's Yokohama Specie Bank (No. 24) and the Banque de L'Indo-Chine (No. 29).

The French bank's showpiece on the Bund was designed in a Classic style by a leading firm, Atkinson & Dallas.

The facade is gray Suzhou granite, the rear is covered with less expensive artificial stone.

Despite its small scale and relatively obscure location, No. 29 maintains harmony and elegance among its more imposing neighbors.

The building is an art piece featuring numerous Greek-style columns. The centerpiece of the facade contains two granite Ionic columns running from the second to third floor. The main entrance is in Doric style, flanked by two columns of polished granite.

Just to embellish the windows, the designer used as many as 22 Doric pilasters, big or small, round or square.

The second floor features a three-sectioned window, with a high, arched central section flanked by two smaller rectangular segments - a classic Palladian window.

The interior of the banking hall is paved with white marble and features six very large Ionic pillars and 12 dark wood French window with pleasing curves. Like other bank buildings on the Bund, this one has a large glass dome over the banking hall; warm light filters through the yellow and opaque white glass.

The French bank ended business in China in 1955 as foreign enterprises retreated from the Chinese market. Renamed the Dongfang Building in 1956, it was used as the Traffic Division of the Shanghai Police Station until 1995.

In 1994 the Shanghai government began moving out state-owned enterprises off the bund to make way for financial institutes and recreate China's Wall Street. No. 29

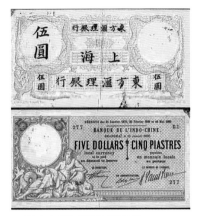

20世纪初东方汇理银行的银行票证
Bank notes issued by the Banque de L'Indo-Chine

was the first to embrace its new role; a branch of China's Everbright Bank opened there in July 1995.

The former owner of No. 29, Banque de L'Indo-Chine, has been renamed Credit Agricole Corp and Investment Bank and returned to China. It set up a branch near the Bund on Yan'an Road in 1991 and maintains the old Chinese name "Dong Fang Hui Li."

In 1997 the bank moved to the city's new Wall Street - Lujiazui in Pudong and operates an office in the World Financial Center. That move to the new financial services area is reminiscent of the decision to build an elegant office on the Bund 100 years ago.

Tips

The ground floor is open to the public. Note the harmonious proportions of the facade and the artful use of columns.

143年前一个春日的午后，外滩33号英领馆新馆竣工。旗帜迎风飘扬，欢庆的人们把香槟洒到奠基石上。英领馆得以落户外滩与首任驻沪总领事巴富尔有关。

卜舫济在1928年出版的《上海简史》中写到巴富尔极具眼光。虽然当时英政府禁止领馆购地建屋，这位总领事却一直在寻觅合适的基地。1846年，他多次考察后终于看中位于苏州河、黄浦江交汇处的名为"李家场"的地块，并自己垫付了部分资金买下这块风水宝地。三年后，英领馆从老城迁到外滩。

虽然地理位置绝佳，但由于建筑质量原因，1849年建成的馆舍在1852年即拆除重建，1870年第二代领馆又不幸毁于大火，两年后又兴建了今日尤存的第三代馆舍。

砖木结构的新馆由英国人格罗斯曼和鲍伊斯在原址设计重建。建筑是维多利亚时代盛期的新文艺复兴风格，面向一个美丽的大花园，底层有连续的券廊，屋顶上还覆盖了中式小青瓦。

"文革"期间英领馆撤离33号，后上海国旅等单位入驻。2003年面积近23万平方米的"外滩源"项目启动，33号变身为"外滩源壹号金融家俱乐部"。

拆除了1万8千平方米的违章建筑后，昔日领馆偌大的花园渐渐恢复生机。园中27棵珍贵古树完好保留，其

中包括苏格兰植物学家福钦赠送的200年树龄的广玉兰。

1843年上海开埠才几天，这位日后从中国带走包括茶叶种子等大量植物样本的福钦先生就到访了。在1847年出版的《漫游华北三年》中，这位和巴富尔一样富有眼光的西方人将上海称为"伟大的门户"："庞大的国内贸易，便利的长江和运河运输，通过此地运送茶叶和丝绸比广州要迅捷得多，而这里又有我国棉纺制造业的广阔市场……毫无疑问，过一两年上海不仅会超过广州，而且会变成一个地位更加显赫的城市。"

权力和财富总是易逝，但建筑与绿树往往长留人间。在33号米色小楼前的花园里，植物学家赠送的广玉兰树青葱依旧。他留在纸上那些关于上海未来的预言，虽已成真，却仿佛还墨迹未干。

参观指南

外滩源1号仅对金融家俱乐部会员开放。路过时可以观察一下昔日英领馆极佳的地理位置，以及园内珍贵的古树。

On a spring midday 142 years ago, the newly completed British Consulate was officially opened at the building that today is No. 33 on the Bund.

The Union Jack was raised, the flagpole was gaily decorated, libations of Champagne were poured on the foundation stones, and a toast was drunk by all present to the success of the new office.

As the last in line of the 23 waterfront heritage buildings on the Bund, No. 33 dates back to the early formation of the Bund. That's why a development project with No. 33 as the centerpiece is named "Waitanyuan," meaning "origin of the Bund."

According to F. L. Hawks Pott's *A Short History of Shanghai* (1928), the first British Consul to Shanghai, George Balfour, was determined to secure a proper site for an office, despite British rules prohibiting consuls from purchasing land in foreign countries. At that time, they were only allowed to work in rented premises.

In April 1846, Balfour finally purchased the property, which was splendidly situated north of the British Settlement. His successor Rutherford Alcock carried on after Balfour resigned five months later and moved the consular office from the Chinese old town to this Bund site in July 1849.

The architectural history of British Consulate on the Bund was filled with twists and turns.

The first consulate building on the Bund collapsed soon after it was built in 1849.

The second one constructed in 1852 was destroyed by fire on December 23, 1870, and many documents were lost. Two years later, the consul hosted a grand ceremony to lay the foundation of the new British Consulate, the present-day No. 33.

At cost of around 6,500 pounds, the new consular office is designed by Grossman & Boyce in neo Renaissance style of the prime Victorian period, featuring an arched loggia as the key element of composition.

The main entrance of the brick-and-timber structure faces south, onto a huge green garden.

The British Consulate retreated from the building during "cultural revolution" (1966–1976). Several state-owned institutions, including China International Travel Service Co Ltd, occupied the building until the Waitanyuan Project kicked off in 2003.

The project covers 229,000 square meters, with its northern end reaching Suzhou Creek. The heritage buildings are mostly sprinkled along Yuanmingyuan Road, Huqiu Road and Beijing Road E. The project was mainly funded by the Huangpu District government and overseas investment.

Waitanyuan is designed to provide services for the financial street on the Bund. Hotels, clubs, retail outlets and apartments will be built inside or outside the dozens of heritage buildings.

Around 18,000 square meters of added buildings were demolished during the renovation to revive the original big lawn of No. 33.

The former British consular office at No. 33 is now a chic venue, which plays a role as a "financiers' club," and renamed "No. 1 Waitanyuan." Both government departments and Shanghai's financiers regularly host conferences and activities here.

A huge sum had been paid just to preserve 27 valuable old trees in the garden, including 200-year-old magnolia trees given to the city by Scottish botanist Robert Fortune. The scientist visited Shanghai in 1843, a few days after the city was declared an open port on November 17. In his 1847 book *Three Years Wanderings in the Northern Provinces of China*, Fortune had called Shanghai "the great gate".

"The large native trade, the convenience of inland transit by means of rivers and canals; the fact that teas and silks can be brought here more readily than to Canton (today's Guangzhou) and, lastly, viewing this place as an immense mart for our cotton manufactures... there can be no doubt that in a few years it will not only rival Canton, but become a place of far greater importance."

Power and wealth come and go, but sometimes buildings and trees survive. Fortune's trees in front of No. 33 are still green, and it seems his ink, writing an accurate forecast of Shanghai, is still wet.

Tips

The former consular office is only open to members of the financiers' club. I would suggest you appreciate the ancient trees and the strategic location of No. 33.

外滩 33 号 2 号楼 No. 33-2

滨水小楼的迷离身世
The Oldest Surviving Waterfront Building

外滩滨水建筑里,到底哪一座历史最悠久呢?最新的研究结果显示,应该是一座文艺复兴风格的灰砖小楼,昔日的英国驻沪总领事官邸,现为瑞士名表百达翡丽的旗舰店。

英国驻沪总领馆原来的馆舍曾毁于1870年的一场大火,1872年重建。在上海建筑史中,1873年竣工的新领馆一直被认为是外滩现存最早的滨水建筑,而官邸要10年后落成。

但根据英国领馆建筑专家马克·博特伦近年的发现,官邸其实建于1871年,比领馆大楼还要早。

马克从大英档案馆的馆藏里找到了相关珍贵史料。主要是1870年前后往来于英国驻华大使、驻沪总领事、国务秘书和领馆工程官员间的通信,内容是商量馆舍建设和使用计划。

这些信件表明,1870年发生大火时官邸已近竣工。这场大火把领馆和大量馆藏文件烧了个精光,但官邸小楼却十分幸运地未受影响。1871年初,原本住在副总领事家中的总领事顺利乔迁新居。

1872年6月3日出版的英文报纸《字林西报》也印证了马克的论断。一篇关于领馆新楼奠基的报道提到,"来自最高法院和领事馆的代表们先在新的总领事官邸集合,准备停当后就一同前往奠基仪式举办的地点。"这表明1872年时官邸已经投入使用。文章还提及仪式结束后人们又回到官邸继续参加午餐会。他们举杯畅饮,发表感言,庆贺新楼奠基。

灰色的官邸风格温馨,与高大气派的英领馆迥然不同。室内设计也舒适实用,底层是起居室、餐厅和书房,二楼有三间卧室。主体建筑外还附带厨房、马厩和洗衣房等,很适合一名19世纪旅沪外交官的家庭生活。

小楼的亮点是细腻的砖饰,由灰、红两色组成,还有拱券和连廊,配以图案精美的彩色地砖,相当耐看。

1949年后小楼曾空置多年,后来被用作幼儿园,也曾经作为多部影视剧的拍摄场景。2012年瑞士名表百达翡丽总裁泰瑞·斯登的母亲格迪·斯登亲自操刀,在保留原始房型结构的基础上,将小楼改造为奢华而不失典雅的旗舰店。这座身世扑朔迷离的外滩小楼,又开始了一段新的历史。

参观指南

建筑作为瑞士名表百达翡丽的源邸对客户开放,有时也会举办历史主题的展览。

The former British Consul General's house in Shanghai, which opened in 2012 as Maison Patek Philippe after renovation, was likely the oldest surviving waterfront building on the Bund.

The first British Consul General in Shanghai, George Balfour, purchased the property splendidly located north of the settlement in 1846 for the erection of a consular office. The first consulate building soon collapsed and the second was destroyed by fire on December 23, 1870. The third one, built in 1872, has survived at No. 1 Waitanyuan.

An array of other consular buildings, including this consul general's house, vice consul general's house and architect's house, were scattered in the compound perching where Suzhou Creek flows into the Huangpu River.

British scholar Mark Bertram's recent study dates the history of the consul general's house to 1871, which was not only older than the central consulate building, but also older than its estimated completion year - 1883, which was widely quoted by Chinese architectural books.

This author of *Room For Diplomacy* mapped an outline of the house's origins from studying archival letters from the 1860s to the 1870s among Sir Rutherford Alcock, British minister in Beijing, Consul

Walter Medhurst in Shanghai, Secretary of State Earl Granville in London and two officials responsible for consular building activities, Major William Crossman and surveyor Robert Boyce.

It was Crossman's recommendation in 1867 for building the consul general's house at the north end of the compound, which was later accepted by London and built by Boyce in 1869.

When the 1870 fire destroyed the central consulate building, the new house was nearly complete and luckily unaffected. The consul, who was living in the vice consul's house to the south of the central building before, took it up as his new residence in early 1871.

The scholar's conclusion is proved by a report in the *North China Daily News* on June 3, 1872, which records the ceremony to lay the foundation stone for the central building:

"The accompany (mainly from HBM Supreme Court and Consulate) assembled in the new house which had been lately built as a residence for the British consul; and adjourned when all was declared ready, to the spot where the ceremony was to be performed."

When the ceremony was completed, "At Mr Boyce's invitation, the party then adjourned to luncheon, which had been laid out in the consular residence."

"After this very pleasant portion of the ceremony was over," attendants proposed toasts and drank to the prosperity of the new consulate.

The detached house was quite a

functional home, featuring a dining room, a drawing room, a study room, a kitchen, a washhouse and a stable on the ground floor and three bedrooms on the second floor. Unlike the central consulate building with high ceilings, the consul-general's house is really refined and fit for a cozy family life in Shanghai.

After 1949, the building remained vacant for years and was later used as a kindergarten attached to a state-owned organization. It may also have been adapted as a set depicting "a Kuomintang prison" for shooting movies themed in old Shanghai.

According to the original interior structure, Gerdi Stern, mother of Patek Philippe president Thierry Stern, decorated the house in a proper, elegant style, turning yesterday's diplomat's home into a stylish flagship of this luxury watch brand from Switzerland.

Tips

The building opens as Maison Patek Philippe which sometimes hosts history-themed exhibitions.

外白渡桥 The Waibaidu Bridge

这是一座长青桥
Shanghai's "Grandma Bridge"

"暮霭挟着薄雾笼罩了外白渡桥的高耸的钢架,电车驶过时,这钢架下横空架挂的电车线时时爆发出几朵碧绿的火花。"外白渡桥充满结构美的身姿出现在茅盾小说《子夜》的开头,也闪动在无数关于上海的作品中。

美丽钢架桥的前身是一座简陋的木桥。1856年,怡和洋行职员威尔斯发现苏州河上摆渡不便,便利用这个商机架设木桥收费,获利甚丰。

后来工部局收购了濒临坍塌的威尔斯桥,新建一座免费通行的木桥。因其临近外滩花园而名为"花园桥"。而这座可以白(免费)渡的桥则被民众昵称为"外白渡桥"。

1907年为便于有轨电车通行,工部局又拆除木桥,新建的铁桥于1908年1月正式通车。桥墩都是用波特兰水泥制成的,基石的材质为大理石。同年出版的英侨杂志《社交上海》详细报道了高质量的新桥:"除了深绿色的栏杆,大桥周身漆成浅灰色。油漆颜料都是英国进口的。"

外白渡桥由于连接外滩与虹口,位置重要,一直车水马龙。1926年的交通流量统计显示,仅5月17日、18日两天早晚7点间,桥上就通过行人50 823人次、人力车14 600辆、汽车4 999辆、公共汽车172辆和有轨电车922辆。

时光流转到2007年,上海市政管理局收到1907年承揽钢结构工程的新加坡厄斯金公司来信,提醒该桥寿命已经百年,需要大修。这封信启动了2009年外白渡桥大修工程,除修复锈蚀部位,还恢复原来的木质人行道,给人走在家里的地板上的感觉,深受好评。

"以今天的技术而言,在苏州河上重建一座现代化的新桥真是太容易了。但这是上海的'外婆桥',见证了这座城市多少年来的欢乐与忧伤。"桥梁专家潘震涛说道。

在《子夜》中,吴老先生从码头上岸后就看到外白渡桥。这位25年未离开故土的老人为避战乱来到上海,坐车离开码头后,他惊讶于窗外的景象:"从桥上向东望,可以看见浦东的洋栈像巨大的怪兽,蹲在暝色中,闪着千百只小眼睛似的灯火。向西望,叫人猛一惊的,是高高地装在一所洋房顶上而且异常庞大的霓虹电管广告,射出火一样的赤光和青燐似的绿焰:Light,Heat,Power!"

历经百年沧桑的外白渡桥今日犹存,专家认为如果按时保养和检验,加强对锈蚀的措施,这座桥至少还可以再使用50年。而它既是历史文物,又具有实用功能,真是一座长青桥。

参观指南

大修后外白渡桥的木板地非常舒服,来来回回地走几遍感觉很好。这座桥在不同天气、不同时间会展示不一样的美,最美还是清晨或黄昏。

"Under a sunset-mottled sky, the towering framework of Garden Bridge was mantled in a gathering mist. Whenever a tram passed over the bridge, the overhead cable suspended below the top of the steel frame threw off bright, greenish sparks."

That was an excerpt from the first paragraph of the famous Chinese novel *Midnight*, written by Mao Dun in the 1930s.

The Garden Bridge, today's Waibaidu Bridge, has been a showpiece of Shanghai since the day it opened in January 1908.

The bridge has played a part in numerous Shanghai-themed books and movies, especially postcards on which locals perform tai chi in front of its double hunchback-shaped silhouette of iron in the morning sunlight.

F.L. Hawks Pott's 1928 book *A Short History of Shanghai* recorded that British merchant Wills organized the Soochow Creek Bridge Company to build the Wills Bridge in 1856, which was "not a very sightly structure" but "the company made a great profit from the tolls collected from those using the bridge." There were complaints from the public about the bridge being a profit-making enterprise.

Finally, the then Shanghai Municipal Council obtained control. They first erected a free bridge by the side of the company's and later purchased Wills' company. As the era of trams arrived and they needed to get across the bridge, the council in 1907 built the steel-frame Garden Bridge, which is what we see today. It is one of the earliest examples of a camelback truss bridge in China.

The English-language *Social Shanghai* paper reported in 1908 that the abutments

有轨电车从外白渡桥上缓缓驶过，20 世纪 20 年代
Trams passed through the Waibaidu Bridge in the 1920s

and pier of the new Garden Bridge were being made with Portland cement and had granite cornerstones.

Perched at such a crucial location, it was always a busy bridge. According to an official 1926 survey, 50,823 people, 14,600 rickshaws, 4,999 cars, 172 buses and 922 trams passed through the bridge from 7am to 7pm on May 17-18.

The bridge also served as a perfect platform to appreciate the beauty of the Bund. Chinese writer Lin Huiyin had written subtle lines about the bridge's night scene, depicting the tiny lights on the bridge as "lotus lights on West Lake on the night of June 18 that gently move in the breeze."

The bridge also witnessed frightening moments. A typhoon on July 27, 1915 killed a person walking across the bridge by dashing him to the ground and blew several people into the Huangpu River.

Later that year, the city's top military officer, Zheng Rucheng, was assassinated by two revolutionaries while his car was driving on the bridge to attend a ceremony at the Japanese Consulate.

In 1907, Messrs Howarth Erskine, Ltd of Singapore obtained the contract for the supply erection of the steel work for the bridge, which was manufactured for them by the Cleveland Bridge & Engineering Company Co of England.

In 2007 Shanghai government received a letter from the century-old contractor reminding it was time for repairing the bridge.

The letter prompted a large-scale repair project in 2009. The bridge was removed and taken to a shipyard in the Pudong New Area for repair. It also needed to be moved for underground work to be done for the Bund Renovation Project before the World Expo 2010 Shanghai.

The old decking and the "character of the bridge" was maintained, with the load transferred from stringer structural elements to the floor beam and then to the main truss. The original curved bracket on the cross frame, which had been straightened during previous renovations, was restored to its nice, old curves. The beloved wooden sidewalk, which had been replaced with concrete, was restored, too.

"With today's technology, it's easy to rebuild a modern, new bridge across the Suzhou Creek. But this is Shanghai's 'grandma bridge,' which had witnessed the sorrow and happiness of the city for so many years," says bridge expert Pan Zhentao, consultant for the project.

"It's a national relic, but it's not only a national relic. It's a bridge that people and cars are still traveling through it. It's an evergreen bridge," Pan says.

In Mao Dun's novel *Midnight*, the Waibaidu Bridge was the first thing that Old Mr Wu saw after he arrived in Shanghai by boat. The old man, who had not left his house in the country for 25 years, was

20 世纪 30 年代外白渡桥一片车水马龙的景象
Perched at crucial locations, it was always a busy bridge in the 1930s

severely shocked by scene in modern metropolitan Shanghai.

"To the west, one saw with a shock of wonder on the roof of a building, a gigantic neon sign in flaming red and phosphorescent green: LIGHT, HEAT, POWER!"

Perturbed by the "light, heat and power" of Shanghai, Old Mr Wu died. But the century-old Waibaidu Bridge still lives, and Pan is optimistic that it will be in use for at least another 50 happy years after the renovation, until 2057, lending its historical perspective to the city.

Tips

It's pleasant to simply walk back and forth on the wooden deck of this old bridge, which shows different kinds of beauty during different hours of the day, especially in the dawn or at dusk.

上海大厦 Broadway Mansions Hotel Shanghai

看见整个外滩的露台
The Screen of the Bund

外滩建筑多设露台，各有风景。但只有上海大厦18楼的露台可同时俯瞰黄浦江两岸的中山东一路和陆家嘴。这个看得见整个外滩风景的露台，由此而著名。

22层高的大楼原名百老汇大厦，是一座近77米高的摩天楼。建筑总面积达31 000平方米，比彼时上海第一高楼国际饭店还要大，远远望去好像外滩的一扇大屏风。

大楼1935年由英资背景的业广地产公司投资建成，集旅馆、公寓功能为一身。当时风行的摩天楼不仅象征权势与成功，也是地价飞涨时代高回报的投资。为完成巨厦，业广地产扩充了建筑部，从英国请来建筑师，还聘请公和洋行作为顾问。

大楼是纯粹的装饰艺术派，屏风造型既充分利用了基地，也有利于房间的朝向和采光，还成为外滩的对景。

20世纪20年代末，发源于法国巴黎的装饰艺术风格开始影响以古典主义为主调的上海建筑风貌。在1929—1938年间，上海建造了38座10层以上的摩天楼，绝大多数都是装饰艺术风。百老汇大厦的立面中央升起，两侧层层跌落，顶部有连续的几何图案，都是装饰艺术派的标志。

摩登大楼建成后出租一空，受到外国记者和洋行雇员的喜爱。可好景不长，日军侵华后租界变为"孤岛"，1939年业广地产低价将大楼转让给日本"恒产株式会社"。二战胜利后蒋介石将大厦作为"励志社"（军官俱乐部）使用。

1949年5月，百老汇大厦这扇"屏风"成为国民党军队抵御解放军的最后一道屏障。

彼得·汤森在1955年出版的《中国凤凰》一书中回忆了这段亲身经历："从百老汇大厦的顶上，能望见硝烟飘到河面上。在西部管制区，可见到烧焦的家园和商店，包括不少外资的不动产。难民涌进城市……走到百老汇大厦的护墙边，一颗子弹冷不丁从头上呼啸而过。我赶紧躲闪，不得不用手和膝盖爬回去。"

新中国成立后大楼更名为"上海大厦"，如今是一家五星级宾馆。东方明珠建成前，到上海大厦登高俯瞰市容是重要的外宾接待节目。上海市外办资深外事工作者夏永芳曾经常陪外宾登临这个"能在几分钟内让他们为上海倾倒"的著名露台。

她说"虽然90年代许多外宾团都安排到东方明珠，可上海大厦仍是我的最爱，因为这座大楼本身就是历史，是东西方文化的结合。从这里可以看见上海的母亲河——苏州河，眺望虹口，欣赏外滩金融街和陆家嘴。虽然经常去，但每次登上这个露台，我仍然激动不已。"

参观指南

建筑作为酒店对外营业。留声机、大堂里的钢琴和英国吧的台球桌都是历史旧物。17楼和18楼的著名露台仅对用餐客人开放。

20世纪30年代刚落成时的百老汇大厦外景
The Broadway Mansions in the 1930s

Broadway Mansions Hotel acts like a huge screen for the Bund from the city's northern territory. As one of the three tallest Bund buildings (the Peace Hotel and the Bank of China are the other two), the 22-floor riverside edifice offers a panorama of the Bund.

The building was built by British firm Shanghai Land Investment Co Ltd in 1930, when Shanghai developers were in the midst of a skyscraper craze to show off the city's modernity, power and success. They were also usually smart investments as land prices skyrocketed during Shanghai's "golden age" of the 1920s–1930s.

It was such a huge project that the total building area amounted to 31,000 square meters, even larger than Park Hotel, the city's tallest building back then. The company had to expand its architecture department and hire architects from the UK to accomplish this giant task.

Perched alongside the Waibaidu Bridge and Suzhou Creek, the steel-structured building was designed as a hotel with some apartments, just like Park Hotel.

The 77-meter-high building is made in the Art Deco style, which spread to Shanghai soon after it first appeared in France in 1925.

The style had a great impact on Shanghai's architectural scene. From 1929 to 1938, Shanghai built 38 buildings taller than 10 floors, most of which were in the Art Deco style.

Designed by British architect B. Fraser, the highrise reveals a four-wing shape to make full use of the land and obtain the best sunlight for hotel rooms.

The facade rises in the center and the cornice near the top is adorned with geological patterns, all very Art Deco. The walls are capped with brown ceramic tiles.

After completion, the occupancy rate was 100 percent as famous international journalists and senior employees of foreign enterprises were quick to rent apartments.

However, after the Japanese invaded Shanghai, the property had to be sold to a Japanese company in 1939 at a price even lower than it cost to build it.

After World War II, Kuomintang leader Chiang Kai-shek reopened the hotel as a guesthouse for Li Zhi She, or The Society for Encouragement, which was like an alumni club for the Huangpu Military Academy. Chiang was president of the academy in the 1920s.

The building's biggest brush with history came in May 1949, when Chiang's soldiers were about to lose to the Communists. Peter Townsend, who had served with the Friends' Ambulance Unit in China, witnessed this scene.

"From the top of Broadway Mansions you could see smoke wreathing upwards across the river. In the western confines you came across charred homes and shops, and a number of foreign-owned properties had been burned down. The refugees poured into the city... when you go out on the parapet of Broadway Mansions a bullet whistles above your head and you duck and crawl back on hands and knees," he wrote in his 1955 book *China Phoenix*.

Shanghai government took over the building after 1949 and renovated the interior. However, a gramophone, a piano, a snooker table and an Ottis elevator survived, all originals from the 1930s.

Of course, the famous 18th-floor balcony is still there. It has hosted a galaxy of honorable guests over the years like late Premier Zhou Enlai and late French President Georges Pompidou.

Kitty Xia, a former official with the Shanghai Foreign Affairs Office, says she has taken numerous foreign guests to the balcony from the 1960s to 1990s.

"The balcony offered the best view of Shanghai," she says. "It was an important part of the itinerary to take foreign delegations to this balcony. Within minutes, our foreign guests were touched by the charisma of Shanghai, such as the Bund, the Huangpu River and Hongkou District where Jewish refugees had sheltered during World War II."

"Although the Oriental Pearl TV Tower was built in 1994 and now offers amazing views, Broadway Mansions is still my favorite. Every time I visit the balcony, I get excited because I love the city so much," she adds.

Today this balcony is probably still the best place to appreciate both the Bund and Lujiazui from one spot.

Tips

You can see the gramophone and the piano at the lobby or play on the historic Snooker pool in the British bar. The famous balcony is open only for diners of the restaurants at the 17th and 18th floors.

浦江饭店　Pujiang Hotel

木香萦绕的老饭店
A Hotel Scented with Oriental Woods

多年以后，海伦·福斯特·斯诺仍旧记得在礼查饭店度过的第一夜。

"我站在这间高耸的客房中间，闻着蚊帐潮湿而发霉的味道。灰色蚊帐笼罩着一张巨大的维多利亚四柱床。"她在自传里（《我的中国岁月》，1984）回忆了那个 1931 年的夏夜。

对于当时到访上海的美国人来说，礼查饭店通常是落脚的第一站。饭店创立于 1846 年，1857 年搬到外白渡桥边，1907 年扩建为今日犹存的新古典主义风格建筑。新店开业，酒店登报做广告自称为"东方的华尔道夫"，是"远东地区最大、最好、最现代的酒店"。

1912 年外侨杂志《社交上海》的一篇报道对比了饭店改造前后的情况："高大的柱饰门厅，酒店的七人乐队为用餐的客人们演奏，制冰设备和电风扇让炎炎夏日美丽而清凉，堪称申城最凉爽的地方之一。老客房里曾经安装着又大又难看的国产金属浴缸，外面漆成棕色，里面泛着绿光。如今换成了修长洁白的英国进口搪瓷浴缸，还有内置的热水系统。"

因为历史悠久，礼查饭店见证了上海开埠以来许多历史性时刻。1846年12月22日，租界工部局的前身——道路与码头委员会在此成立。酒店也是上海最早使用煤气、点亮电灯、用上自来水、提供电话和出租车服务的酒店。在鼎盛时期，礼查饭店是外国贵宾旅沪下榻的首选。

此外，饭店还是美国报业著名的"密勒氏王朝"所在地。《密勒氏评论》由《纽约先驱论坛报》记者密勒于1917年在上海创办，是远东地区最有影响力的政治周刊。

密勒先生长住礼查饭店，他的两位同业后辈约翰·鲍威尔和后来写出《西行漫记》的埃德加·斯诺也先后在此长住。海伦是经密勒介绍后入住的。她抵沪首日就结识了斯诺，第二年在外滩答应了斯诺的求婚。

"美国人习惯下榻这家古老的礼查饭店，而不是那家新建的更昂贵的华懋饭店。"海伦在自传中写道。

如今，饭店保留了海伦入住时期的格调。一楼孔雀厅悬挂着水晶吊灯，墙面镶嵌着无数面镜子，厅内还矗立着多根大理石柱子。镶木地板形成巨大的螺旋图案，看久了眼前仿佛闪现旋转的裙裾。

上海是个"不夜城"。1934年的英文导览中介绍了这座城市"燃烧的夜晚"——夜生活从下午茶舞开始，持续到凌晨2点或早餐前的任意时间才结束。外滩三家酒店各有特色："华懋饭店有茶舞和晚餐舞会，夜总会节目是精心制作的。礼查饭店的冬季茶舞和古典音乐会很受欢迎，而汇中饭店则在下午茶和晚餐时间有音乐会。"

1954年饭店收归国有，1959年更名为"浦江饭店"，用作政府接待。1990年上海证券交易所开市第一槌响彻孔雀厅，老饭店又一次见证了历史性时刻。1998年酒店改为青年旅馆，直到2002年才整修为星级酒店，吸引了很多怀旧的客人。

在上海的第一夜，海伦把酒店房间里厚重的木抽屉全部打开。"抽屉内侧镶着樟木或檀香木，还有的是肉桂木。它们的香气与陈旧味道的空气搅在一起。后来无论何时只要闻到这种味道，我就对东方的日子无比怀念。"

参观指南

三楼有展示酒店历史的展览和名人住过的房间。孔雀厅地板上那美丽的螺旋图案也是一定要看的。

American author Helen Foster Snow remembered the moment she caught the treaty-port mystique in the Astor House (today's Pujiang Hotel) when she arrived in Shanghai in the summer of 1931.

"I stood in the middle of my big, lofty room (costing about two US dollars a day, with food) and sniffed the dampish, gray-mildewed mosquito net covering the whole of the huge Victorian four-poster bed," she wrote in her 1984 memoir *My China Years*.

For many Americans arriving in Shanghai before the 1940s - like Helen who later married *Red Star Over China* author Edgar P. Snow - the Astor House was a first stop.

Originating from a boarding house known as the Richard's Hotel in 1846, this leading hostelry was relocated in 1857 to its current location near the Waibaidu Bridge and renamed Astor House.

Today the hotel is comprised of a six-story and a four-story buildings in Neo Classical style, designed by Davies & Brooke Architects and expanded in 1907. After the grand expansion, the hotel advertised frequently on leading English newspapers, proudly calling itself the "Waldorf-Astoria of the Orient" and the "largest, best and most modern hotel in the Far East."

The 1912 edition of the English-language magazine *Social Shanghai* published an interesting article about the new "luxurious" hotel compared with its merely "comfortable" predecessor back in 1896.

The article admired the elaborate dining halls, the lofty pillared foyer, the seven-piece hotel band playing at every meal, an ice-making plant, plus the ample supply of electric fans which keep the hotel "so beautifully cool in summer."

Even the bathrooms were compared. The old bathrooms were fitted with big, ugly native bath tubs made of metal, which was grained brown outside and enameled with a green glaze inside. The latest bathrooms have long white enameled baths, which have been imported from the UK with a hot water system laid inside.

Among Shanghai's historic hotels, Astor House boasts the longest history and witnessed major events since the city's port opened in 1843. On December 22, 1846, a meeting to establish the Committee on Roads and Jetties was hosted in the hotel, still known as Richard's Hotel. The committee, which organized construction of roads, jetties and bridges, was later regarded as the cradle of the Shanghai Municipal Council that practically administered the International Settlement until the 1940s.

The Astor House enjoyed many firsts as our Paris-of-the-Orient city imported Western inventions. It was the city's first hotel to use coal gas, electric lamps and tap water, and to provide telephone and taxi services. During its best days, Astor House attracted the majority of foreign guests, including Bertrand Russell and Charlie Chaplin, to name a few.

It was also the seat of a journalism dynasty - the Millard-Powell-Snow succession - referring to Thomas F. Millard, John B. Powell and Edgar P. Snow, three graduates

of the University of Missouri School of Journalism.

Milliard, widely considered "the dean of all newspapermen in the East," had lived in the hotel after arriving in 1900 to cover the Boxer Rebellion for the New York Herald. His proteges Powell and Snow resided in the hotel afterwards.

With an introduction from her father's Stanford University alumni friends to Milliard, Helen Foster also checked in the hotel soon after she landed in Shanghai.

"It was in the American tradition to go not to the expensive new Cathay Hotel, but to the ancient Astor House," she recalled in her memoir.

Today the decorations of the "golden" 1930s that Helen Foster knew are stunningly preserved on a large scale. Stepping into the famous pillared foyer, visitors feel they are stepping into another era.

The highlight is the two-story-high Peacock's Hall on the ground floor. It's adorned by grand domes, crystal chandeliers, mirrors and marble pilasters. The centerpiece is the original parquetry flooring designed in a pattern of huge spirals, giving an artistic feeling of flowing and spinning. Imagine the beautiful lacy skirts swirling across this floor a century ago.

Shanghai used to be a city of blazing night. According to the 1934–35 Standard Guide Book *All About Shanghai and Environs*, night life began with the tea-cocktail hour and ended at anytime from 2am until breakfast. Three hotels (Astor, Cathay, Palace) on the Bund were all popular.

"Shanghai likes to dance and to be entertained. The Cathay Hotel features tea dances and dinner dances, with elaborate floor shows. During the winter season the Astor House's tea dances and classical concerts are popular. The Palace Hotel offers concerts during the tea and dinner hours."

Astor House was taken over by the Shanghai government in 1954 and used as office for a state-owned trading company. In 1959 it was renamed the Pujiang Hotel and became a government guesthouse.

In 1990 the hotel embraced a new role - the Shanghai Stock Exchange was launched with a ceremony in the Peacock Hall on December 19, 1990. It witnessed the birth of the securities market in the new China.

Then the former luxury hotel was transformed into a hostel with dormitories for foreign backpackers before it was renovated to a three-star hotel after the 2002 renovation.

Indeed, the hotel is probably the best place to get a feel of the times when the nights were ablaze. Perhaps nowhere in the Bund area is there a longer and more intense history.

On her first night in this Oriental hotel, Helen Foster struggled to open heavy mahogany or teakwood drawers.

"They were lined with camphor wood or sandal wood, some with cassia wood, and their perfume mingled with the musty air. Whenever I chance upon that scent, I feel nostalgic for the East," she wrote.

Tips

I suggest visiting the central hall on the third floor that is lined with celebrities' rooms and features an exhibition about the hotel's history. And of course, appreciate the spiral patterns on the floor of the stunningly beautiful Peacock Hall.

俄罗斯联邦驻上海总领事馆
Consulate General of the Russian Federation in Shanghai

反复开关的领事馆
Openings and Closings

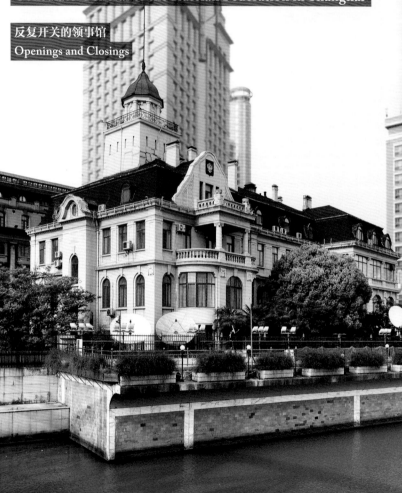

上海众多领事馆中，俄罗斯领事馆的命运最具戏剧性。虽然这座红瓦灰墙的领馆大楼自建成起一直是俄领馆，却开开关关，历经沧桑。

大楼由德国建筑师里约伯设计，1917年建成。整座建筑融会了巴洛克和德国复兴主义元素，是俄罗斯最美丽的驻外使领馆之一。

俄国1896年即在上海设立领馆，但最初的馆舍分散各处。1911年新上任的总领事格罗瑟富有远见，他向沙皇陈述上海的重要地位，从而成功获得建馆专款。

世事难料，新馆开张后不到一年就因"十月革命"爆发而关闭，直到1924年中苏建交后作为苏联领馆重新开张。但1927年领馆又因局势不稳闭馆，1932年重开，1941年太平洋战争爆发后再次关闭，1949年又在原址开馆。

至此，这一出反复开关的戏剧还未落幕。灰色大楼在中苏关系恶化后于1962年再次闭馆，到1986年复馆，1991年苏联解体后大楼更名为俄罗斯联邦总领馆，直至今日。

政治的动荡不仅让领馆开关个不停，也影响了俄人的命运，许多人被迫逃亡上海。根据美国学者罗茨·墨菲的统计，1936年在沪俄侨有15 000名，数量仅次于日本侨民。俄国侨民像潮水一样涌进无需签证的上海，包括乞丐、贵族和大量士兵。

美国著名报人约翰·鲍威尔恰好在俄领馆落成的1917年来到上海，他在回忆录中提到俄国人的快速适应能力。"士兵找到中国银行保安或守夜人的工作，俄国女人开了服装店、帽子店和美容沙龙，而在法租界几乎每个街区都出现了俄式餐厅。"他认为当时俄侨在上海扮演的角色很重要，其社会阶层正好介于西方白领和中国苦力之间。

如今俄领馆一楼是签证处，办公在二层、三、四层为工作人员居住。面向浦江饭店的雕花大门上高悬俄罗斯国徽——以金色双头鹰为图案。门厅地面由马赛克铺就，别致的图案十分少见。小巧的门厅还装点着黑红色大理石柱子、镜子和螺旋形乌木楼梯。会议厅的护墙裙也是乌木的，视觉焦点是一个极具奢华感的铜饰大理石壁炉。

领馆还有面向外白渡桥的露台，江景迷人。站在露台上发现楼前居然还有一个精巧的花园，布置着迷你池塘和六角凉亭。

作为一座历史建筑，这栋灰色大楼得以完整留存，并保持百年前初建时的功能，可以说是非常幸运。虽然，这扇雕花大门曾经开了又关，关了还开。

参观指南

领馆只有签证区对外开放。在外白渡桥上欣赏俄领馆的外观是一个很好的角度，同时还可以享受徐徐的江风。

Among Shanghai's rainbow of consulate buildings, the gray, red-roofed Russian Consulate General fronting the Huangpu River is one of the most dramatic.

Though it was built for and is still used as the Russian Consulate General in Shanghai, the architectural history of the grand building has been filled with twists and turns.

Perching at the confluence of the Huangpu River and Suzhou Creek, the consulate can be appreciated from the Waibaidu Bridge. Red roofing tiles contrast with the pale gray facade, which has a solid, severe look.

It was erected in 1916, designed by German architect Hans E. Lieb who adapted Baroque and German Renaissance elements for the 3,264-square-meter structure.

The first Shanghai consulate of the Russian Empire was established in 1896, however, at that time the buildings were sprinkled in different parts of the city.

Soon after his appointment in 1911, the new Consul General Victor Grosse reported to the Czar on Shanghai's geographic and economical importance. The Czar approved Grosse's plans for Shanghai, one of which was to build the new consulate building.

The four-story structure was built in 1914, completed in 1916 and formally opened in January 1917. It was one of the most beautiful Russian consular buildings in all countries.

Only months after the grand opening, however, the 1917 Russian Revolution broke out and all Russian embassies and consulates overseas were closed, including the one in Shanghai.

In 1924, the Shanghai consulate of former Soviet Union reopened but closed in 1927 because of insecurity in the city controlled by the Kuomintang regime.

The Russian consulate opened again in 1932 after the ruling Kuomintang and former Soviet Union governments restored diplomatic relations. But it closed again in 1941, around the time of the Japanese occupation of Shanghai. It reopened in 1949 when the People's Republic of China was established and new diplomatic ties were forged.

That's not the end of the open-and-close drama. In the late 1950s, Sino-Soviet relations deteriorated and also Soviet experts assisting in China were ordered home in 1962. On September 28, the consulate closed again and did not reopen until 1986, when relations improved.

Following the dissolution of the former Soviet Union, the building was renamed the Russian Consulate General in Shanghai at the end of 1991.

Russia's political changes not only affected the consular building, but also the tens of thousands of Russian refugees early last century.

Russians constituted the city's largest European population. In 1936 Russian population in Shanghai reached 15,000.

Russian refugees, including many soldiers, flooded into visa-free Shanghai on almost every train and ship from the north. They ranged from beggars to the former aristocracy.

Russians acted quickly to gain a foothold in the city, wrote American Journalist John Powell, who arrived in Shanghai in 1917.

Former Cossack soldiers became bodyguards for rich Chinese merchants and night watchmen at banks and business houses. Russian women of some means opened fashionable clothing shops, millinery shops and beauty parlors.

There were one or two Russian restaurants seemingly on every block, particularly in the former French concession where the majority of Russians resided.

"The Russians filled an important niche in the city, occupying a position between the normal white-collar Occidental population and the Chinese who did all the work, Powell wrote.

Today the building's first floor is visa section, the second floor contains offices and the residence of the consul general. Some employees live on the upper two floors.

Behind a sculptured wooden gate graced by the Russian imperial double-headed eagle, pattern of the country's national emblem, the interior looks more beautiful and far more opulent than imagination.

The vestibule has a mosaic floor in

a delicate pattern, a dark-wood spiral staircase, black and red marble columns, a large mirror and dark wainscotting. The centerpiece of a meeting hall is large, richly decorated fireplace. Outside is a mini garden with a lovely pond and a pavilion.

Ups and downs, openings and closings. Compared with other heritage Bund buildings, the red-roofed, pale gray consulate is rare and fortunate in retaining its original function.

Tips

Except for the visa section, the building is closed to the public. However, you can appreciate the facades from the Waibaidu Bridge while enjoying the river breeze.

黄浦公园 The Huangpu Park

美丽外滩的过去和未来
All for a Better Bund

一本1902年出版的英文书《上海的日日夜夜》里写道，有人问一位刚来上海的英国传教士，租界里什么最吸引人，他回答"是公家花园里的孩子们"。

公家花园就是外滩的黄浦公园。孩子们欢笑嬉戏的公园确实是一道很动人的风景。

这块地由黄浦江和苏州河冲积的泥沙淤积形成。工部局听取了工程师克拉克的建议，出资将浅滩填平，慢慢地建起一座美丽的公园。

这个精致的英式花园设计得很特别，中央绿地由篱笆和鲜花环绕，又种植了乔木和灌木，宁静怡人。

《上海的日日夜夜》一书中还提到公园的植物："在这里你会看到最早绽放的雪莲、最美的风信子和最丰富多彩的郁金香，还有最好的玫瑰和当季最棒的植物……很多人不知道有些植物都是进口的，如英国的七叶树和榛树，当然最壮美的要数'规矩会堂'前的木兰。"

植物营造了迷人的氛围，但公园的灵魂却在音乐。1881年工部局成立了一支交响乐队，在公家花园的木制凉亭定期演奏。音乐会不售门票，但有偿出租帆布椅给听众。演出大受欢迎，最初的600张椅子后来增加到1 200张。

时至今日，关于这个公园的话题还是围绕其禁止华人入园的规定，而美丽公园和外滩背后的故事却鲜为人知。

1869年，美国商人金能亨卸任工部局总董后从日本横滨写来长信，强烈反对继任者将滨江大道用于建造码头仓库的计划。他的来信多少改变了外滩滨江大道沦为码头的命运。

141年后，2010年世博会前，上海下决心拆除了设计寿命100年、服役仅11年的延安路高架下匝道，将过境交通引入地下，把地面更多地还给漫步外滩的人们。

《城记》作者王军称，外滩改造工程是著名的波士顿"大开挖"在中国的翻版。虽然外滩工程规模小得多，但两者均位处市中心的滨水景观区，均是拆除高架桥，修复地面步行系统，回归人的尺度。

如今，大幅拓宽的外滩滨水长廊与外滩公园蜿蜒衔接，漫步外滩的感觉好极了。

在1902年的这本英文著作中，一位匿名作者描绘了一个外滩公园音乐会的夜晚。"这个东方夏日的晚上，星光熠熠的苍穹下，音乐家魔力指挥棒的挥舞间，所有关于过去、现在和未来的那些焦虑、担心、悲伤和害怕，在这短短两小时里都烟消云散了。"

这位感性的作家也许还不了解，这个美丽星夜的过去和未来，人们为了更美好的外滩而付出的努力和辛劳。

参观指南

在黄浦公园和宽敞的滨江大道上徜徉时，可以感怀一下历史，这美丽的外滩确实来之不易。

上图：洋人在公家花园散步　下图：19世纪90年代的公家花园音乐亭
Above: People wandered in the Public Garden　Below: Shanghai Public Orchestra performed concerts(1890s)

In the 1902 book *Shanghai By Night and Day*, a British missionary new to the city was asked what struck him as being most worthy of notice in the foreign settlement. He said, "the children in the Public Garden."

Now known as Huangpu Park, this little oasis is indeed a joyous sight whenever children are seen playing, laughing and running about.

F. L. Hawks Pott's 1928 book, *A Short History of Shanghai*, reveals how the Public Garden became a place for leisure and recreation.

"Originally the land it now occupies was known as the Consular Mud-flat, being formed by the silt deposited by the meeting of the waters of the Whangpoo (Huangpu) River and Soochow (Suzhou) Creek."

At the advice of engineer J. Clark, a member of the Shanghai Municipal Council at the time, and at considerable expense, the land was filled in. A beautiful park was developed gradually out of what had been an unsightly mud flat.

It was a small but well-designed English garden, with nice plants, delicate architectural details and thoughtful facilities. The central lawn had fences and flowers surrounding it. Tall trees and thick shrubs were also planted to make the garden more secluded.

"In it you will see the earliest of snowdrops, the most beautiful of hyacinths, the richest of tulips, the finest of roses and the best of everything else in season... Numbers of people don't know that there are plants in the garden that are not indigenous to this part of the world. There is a horse-chestnut for example from home, and a bunch of hazels, but the most magnificent trees are the beautiful magnolias in front of the Masonic Hall," describes the 1902 book.

While plants provided a leisurely setting, music was the park's soul. In 1881, the Municipal Council founded Shanghai Public Orchestra, which performed concerts in a wooden pavilion in the garden, as well as in other public parks around the city.

The orchestra didn't sell tickets, but rented out canvas chairs to those who wanted to watch the performances. Numbers grew so much that the orchestra committee had to provide more than 1,200 chairs instead of the initial 600.

It's lesser known that various expatriates contributed to the Bund's development and beautiful scenery. Aside from Clark, Robert Hart, inspector-general of China's Imperial Maritime Custom Service, the Public Orchestra's Italian conductor Maro Paci and especially American merchant/vice consul Edward Cunningham, all left lasting footprints on the Bund.

In 1869, Cunningham wrote a letter to the Shanghai Municipal Council from Yokohama, Japan. He strongly opposed the use of the river frontage from Yang King Pang (today's Yan'an Road E.) to the Public Garden for ship wharves. The council changed its decision at his advice and the promenade was saved.

About 141 years after Cunningham had written that letter, Shanghai completed its 3-year revamp of the Bund in 2010.

The elevated highway to the Bund

was demolished decades earlier than its designed service time. A tunnel was built beneath the Bund to reduce the number of lanes from 10 to 4, which made the Bund more "human" for pedestrians and greatly increased the space for public activities.

Renowned Beijing journalist Wang Jun compared the Bund project with Boston's "Big Dig". Both projects were located in the city center along the waterfront, both demolished an elevated highway, built underground tunnels and placed an emphasis on people.

Shanghai Tongji University Vice President Wu Jiang, who participated in the project, says China's urban projects had previously paid more attention to economic interests than returning space to the people. But the Bund revamp showed a spirit to care more about people and public life instead of only GDP.

Now the area is linked with Huangpu Park via the promenade.

In the 1902 book, the anonymous writer described a band playing in Public Garden.

"Therein under the starry canopy of an Eastern summer night, which of itself alone is something to live for, the worries, the cares, the sorrows, and the fears of the past, the present, and the future are for two brief hours, under the magic wand of the musician, wafted to the limbo of forgetfulness."

The writer probably didn't know how many people or how much effort went into making the Bund what it was and still is today. Nevertheless, it's still there for all to enjoy.

Tips

Wander on the spacious waterfront promenade while contemplating the city's history.

拼图外滩

对外滩的兴趣曾经一般，人多闹哄哄，留给游客们拍照吧。后来写外滩专栏，一座楼、一座楼地写，仿佛带领读者游历，写到哪小红旗插到哪。以白纸一张的心态探索外滩，效果反而好，而我也被外滩征服了。这段日子，虽然风吹日晒、疲惫辛苦，但是可以把一个个外滩建筑拼成一幅风云变幻的美图，真是件有意义又有趣的事。

曾经担心23幢临江建筑"良莠不齐"，因为很多书对于外滩名楼（如和平饭店）的介绍常洋洋洒洒好几页，但"冷门建筑"（如总工会大楼）仅有几段介绍，点到为止。为此我特意准备了讲述外滩建筑演变史的副栏，以填补"冷门建筑"的版面。谁知一直写到最后，这个副栏都没有用上，只因中山东一路上的大楼，细细挖掘个个精彩，没有"冷门"。

而这些风格迥异的大楼组合在一起，无意间形成一道迷人的风景线。不仅今天的我们体察到外滩和谐的整体美，1927年当今日外滩已基本成型时，英文《远东时报》的记者也如此写道："这些风格多样的建筑由来自不同国家的能工巧匠设计，而工部局从未要求过以某种风格来和谐统一。也许是巧合，这么多种风格建筑配在一起的效果竟然相当悦目。近年新建的横滨正金银行、台湾银行、字林西报大楼、汇丰银行、海关大楼和沙逊大厦都极好地融入其中。"

徐家汇藏书楼的墨绿色台灯前，我读到泛黄报纸上的这段文字，非常激动。迷人的外滩是上海城市发展的必然，但可能更是一个巧合的礼物。生活在这座城市里的人，应该体会和珍惜这份幸运。

20世纪30年代在外滩工作、生活的美国女记者项美丽在自传里写道："上海一直在不停地变化。"在"拼图"过程中，我也深深感到从西人看中外滩的那一天起，驱动上海不停变化的动力就没有变，那是很多很多的梦想。

这本书中提到不少到外滩来寻找机遇的传奇历史人物，有总领事、传教士、商人、官员、建筑师和女作家等。年富力强的他们在梦想和欲望的驱动下，把泥滩建设成金滩。而今天我们这座城市，不依然是一个盛满梦想的巨大容器吗？在这里，一幅更大的外滩之图或许正在拼贴中。

乔争月
2015年5月

Making a New Bund Puzzle

Living in a quiet lane of Wukang Road, I was not a big fan of the Bund which had always been regarded as a noisy attraction for tourists. But after this full exploration, I had been conquered by the Bund.

Worrying that some not-so-well-known Bund buildings (such as the youngest No. 14) might lack interesting stories, I had prepared a sidebar about "architectural evolution on the Bund" to fill the pages. To my surprise, the sidebar had never been used at all. Each building on the Bund, if you dig deep, has a lot to tell. And all of them mingle together in a harmonious way. Not only do we feel so, a journalist wrote the same in 1927 on the English newspaper Far Eastern Review, when the Bund that we see today was almost in shape.

"They have been designed by numbers of craftsmen of different nationalities and concepts of style, and at no time has the Municipality made an attempt to enforce harmony in any locality. It is partly by chance, therefore, that the diversity offered to the visitor here, is so pleasing to the eye, and that the recently erected Yokohama Specie Bank, Taiwan Bank, North-China Daily News and Hongkong and Shanghai Bank Buildings, the new Customs Building and Sassoon House, will blend so admirably."

I was very excited when I read these words from the yellowish old newspaper at the Xujiahui Library, a branch of Shanghai Library specializing in the collection of old Shanghai archives in foreign languages. The beautiful Bund was a natural consequence of the city's urban development, but it was probably a gift by chance. We who live and work in Shanghai should indeed cherish her.

Writing the Bund column from No. 1 to No. 33, I felt like playing a beautiful puzzle on the Bund. Piece by piece, finally I have managed to get a full picture of the Bund and Shanghai.

Those mentioned in this book, diplomats, missionaries, businessmen, architects and writers etc, they came to the Bund with ambitions or dreams and together they turned the mud land into a billion-dollar skyline. American writer Emily Hahn wrote in her biography that "Shanghai is always changing." And I believe what has kept Shanghai always changing since 1843 has never changed. Our city is still a container of a lot of dreams and desires.

And every one of us is now making a larger, more stunning Bund puzzle bit by bit.

Michelle Qiao
May 2015

推荐阅读
Recommended Readings

一、专著

1 伍江. 上海百年建筑史: 1840—1949（第二版）. 上海: 同济大学出版社, 2008.
2 钱宗灏, 等. 百年回望——上海外滩建筑与景观的历史变迁. 上海: 上海科学技术出版社, 2005.
3 郑时龄. 上海近代建筑风格. 上海: 上海教育出版社, 1999.
4 罗小未. 上海建筑指南. 上海: 上海人民美术出版社, 1996.
5 陈从周, 章明. 上海近代建筑史稿. 上海: 上海三联书店, 1988.
6 常青. 摩登上海的象征——沙逊大厦建筑实录与研究. 上海: 上海锦绣文章出版社, 2011.
7 常青. 都市遗产的保护与再生——聚焦外滩. 上海: 同济大学出版社, 2009.
8 常青. 大都市从这里开始——上海南京路外滩段研究. 上海: 同济大学出版社, 2005.
9 张姚俊. 外滩传奇. 上海: 上海文化出版社, 2005.
10 上海章明建筑设计事务所. 上海外滩历史建筑（一期）. 上海: 上海远东出版社, 2007.
11 唐玉恩. 上海外滩东风饭店保护与利用. 北京: 中国建筑工业出版社, 2013.
12 王方. "外滩源"研究: 上海原英领馆街区及其建筑的时空变迁（1843—1937）. 南京: 东南大学出版, 2011.
13 李欧梵. 上海摩登: 一种新都市文化在中国（1930—1945）（修订版）. 毛尖, 译. 北京: 人民文学出版社, 2010.
14 许乙弘. Art Deco 的源与流: 中西摩登建筑关系研究. 南京: 东南大学出版, 2006.
15 上海锦绣文章出版社, 等. 外滩 12 号. 上海: 上海锦绣文章出版社, 2007.
16 上海锦绣文章出版社, 等. 外滩 20 号. 上海: 上海锦绣文章出版社, 2011.
17 Johnston T, Erh D. A Last look: Western Architecture in Old Shangahi. Old China Hand Press, 1993.
18 Pott F F. A Short History of Shanghai. D.D China Intercontinental Press, 2008.
19 Hibbard P. The Bund. Odyssey Books & Guides, 2011.
20 Snow H F. My China Years. William Morrow & Co, 1984.
21 Hahn E. China to Me a partial biography. 1944.
22 Barber N. The Fall of Shanghai. Coward, McCann & Geoghegan, 1979.
23 Sergeant H. Shanghai. Jonathan Cape Ltd, 1990.
24 Ken Cuthbertson. Nobody said not to go. Faber and Faber, Inc, 1998.

二、近代英文报刊

1 《远东时报》（Far Eastern Review）
2 《字林西报》及其周末版《北华捷报》（North China Daily News & North China Herald）
3 《大陆报》（The China Press）
4 《密勒氏评论》（Milliard's Review）
5 《社交上海》（Social Shanghai）
6 《上海泰晤士报》及周日版（Shanghai Times & Shanghai Sunday Times）

图片来源

张雪飞
P10-11、P13、P14、P16、P21、P23、P24、
P29-30、P32、P36、P40、P42、P43、P46、
P48、P49、P51、P52、P54、P55、P56、P58、
P59、P60、P61、P64、P68、P70、P74、P76、
P77、P80、P82、P84、P86、P88、P90、P91、
P94、P96、P97、P100、P101、P102、P104、
P106、P108、P109、P110、P114、P118、P120、
P122、P123、P124、P125、P128、P130、P132、
P134、P136、P138、P144、P146、P147、P148、
P152、P156、P158、P160、P162、P164、P165、
P166、P170、P172-173、P176

乔争月
P28

华尔道夫酒店
P45

外滩源1号
P140、P141、P142

上海市档案馆、上海市图书馆、上海市城市
建设档案馆、黄浦区人民政府档案馆
P2-3、P4-5、P6-7、P8-9、P34、P38、P39、
P44、P45、P62、P67、P72、P73、P79、P93、
P98、P117、P127、P139、P150、P151、P168

上海大厦
P154

推荐阅读-24
p111

Image Source

Zhang Xuefei
P10-11, P13, P14, P16, P21, P23, P24, P29-30, P32,
P36, P40, P42, P43, P46, P48, P49, P51, P52, P54,
P55, P56, P58, P59, P60, P61, P64, P68, P70, P74,
P76, P77, P80, P82, P84, P86, P88, P90, P91, P94,
P96, P97, P100, P101, P102, P104, P106, P108, P109,
P110, P114, P118, P120, P122, P123, P124, P125,
P128, P130, P132, P134, P136, P138, P144, P146,
P147, P148, P152, P156, P158, P160, P162, P164,
P165, P166, P170, P172-173, P176

Michelle Qiao
P28

Waldorf Astoria Shanghai
P45

No. 1 Waitanyuan
P140, P141, P142

Shanghai Municipal Archives, Shanghai Library,
Shanghai Urban Construction Archives,
Huangpu District Archives Bureau
P2-3, P4-5, P6-7, P8-9, P34, P38, P39, P44, P45, P62,
P67, P72, P73, P79, P93, P98, P117, P127, P139,
P150, P151, P168

Broadway Mansions Hotel
P154

Recommended Readings-24
P111

图书在版编目（CIP）数据

上海外滩建筑地图/乔争月等著.--上海：同济大学出版社，2015.11（2024.5重印）

ISBN 978-7-5608-5867-8

Ⅰ.①上 Ⅱ.①乔 Ⅲ.①建筑物-介绍-上海市 Ⅳ.①TU-862

中国版本图书馆CIP数据核字(2015)第128501号

上海外滩建筑地图

乔争月　张雪飞　著

责任编辑：常科实
助理编辑：孙彬
责任校对：徐春莲
装帧设计：孙晓悦
出版发行：同济大学出版社 www.tongjipress.com.cn
地　　址：上海市四平路1239号 邮编：200092
电　　话：021-65985622
经　　销：全国新华书店
印　　刷：上海雅昌艺术印刷有限公司
开　　本：787mm×1 092mm 1/36
印　　张：5
印　　数：23701—26800
字　　数：149 000
版　　次：2015年11月第1版
印　　次：2024年5月第8次印刷
书　　号：ISBN 978-7-5608-5867-8
定　　价：48.00元